The Crash
of Piedmont Airlines
Flight 22

The Crash of Piedmont Airlines Flight 22

Completing the Record of the 1967 Midair Collision Near Hendersonville, North Carolina

Paul D. Houle

McFarland & Company, Inc., Publishers
Jefferson, North Carolina

LIBRARY OF CONGRESS CATALOGUING-IN-PUBLICATION DATA [new form]
Names: Houle, Paul D., 1964– author.
Title: The crash of Piedmont Airlines Flight 22 : completing the record of the 1967
    midair collision near Hendersonville, North Carolina / Paul D. Houle.
Description: Jefferson, North Carolina : McFarland & Company, Inc.,
    Publishers, 2016. | Includes bibliographical references and index.
Identifiers: LCCN 2015046926 | ISBN 9781476662244 (softcover : acid free paper)
Subjects: LCSH: Piedmont Airlines Flight 22 Crash, N.C., 1967. | Aircraft
    accidents—North Carolina—Hendersonville Region. | Aircraft accidents—
    Investigation—United States.
Classification: LCC TL553.525.N8 H68 2016 | DDC 363.12/46509756532—dc23
LC record available at http://lccn.loc.gov/2015046926

BRITISH LIBRARY CATALOGUING DATA ARE AVAILABLE

ISBN (print) 978-1-4766-6224-4
ISBN (ebook) 978-1-4766-2252-1

© 2016 Paul D. Houle. All rights reserved

No part of this book may be reproduced or transmitted in any form
or by any means, electronic or mechanical, including photocopying
or recording, or by any information storage and retrieval system,
without permission in writing from the publisher.

Cover image © 2016 Thinkstock

Printed in the United States of America

McFarland & Company, Inc., Publishers
  Box 611, Jefferson, North Carolina 28640
  www.mcfarlandpub.com

To all of those who lost their lives
in the sky above Hendersonville, North Carolina,
on July 19, 1967. The truth is
the most beautiful memorial of all.

"We are buying jets that haven't been fully designed and with millions of dollars we don't have. We are going to operate them off airports that are too small, into an air traffic control system that is too slow, and we must fill them with more passengers than we have ever before."—C. E. Woolman, president of Delta Airlines, 1957

Table of Contents

Acknowledgments — ix
Preface — 1
Introduction — 4

1. "I have a flight plan for you" — 7
2. Passenger List — 17
3. "Twenty-two is ready to go" — 21
4. "Twenty-two rolling" — 31
5. "Somebody got an ashtray on fire?" — 45
6. "A large ball of smoke and flame" — 50
7. "We think it might be a big one" — 58
8. "Deliberate, continuous leaks" — 68
9. "This process is conducted entirely in the open" — 80
10. "There is nothing in the budget request for radar systems" — 87
11. "I really think you have been derelict in your duty" — 102
12. "A compromise of aviation safety" — 110
13. Reaching a "Breakdown" Point — 115
14. "Every major FAA air traffic control facility is short of personnel" — 122
15. "A family affair" — 132
16. "Erratic speed control during enroute climb" — 136

17. "That's just the cigarette that's on fire"	145
18. "We'll turn off, go direct to the VOR"	152
19. "I did not participate in the NTSB's investigation"	156
20. "Since we are in a real budget squeeze, aviation must take its lumps"	162
21. "We, the public, prefer simple answers"	166
22. "Poor cockpit discipline"	179
23. "At the very least, the controller should have requested a read-back"	182
Chapter Notes	187
Bibliography	203
Index	209

Acknowledgments

The author would like to thank the following individuals for their help on this project. Without the right people in the right places supplying me with the right information, this book would never have been completed. David Carroll of Charleston, SC; Walt Wooton of Spartanburg, SC; Bill Sudduth, head archivist at the University of South Carolina, Columbia; Melissa Gottwald, archivist/special collections librarian, Embry-Riddle Aeronautical University, Prescott, AZ; Liza Talbot, archivist, Lyndon Johnson Presidential Library; Drew Levinson of Drew Levinson Media; Mary Evelyn Tomlin, programs specialist, National Archives and Records Administration. Particular thanks go to Peter J. Houle of New York City, Andy Hayden of Campobello, SC, Eddie King of Campobello, SC, and AJ Viviano of Inman, SC. A special thanks also to Susan Snowden of Snowden Editorial Services.

There are many others who helped with this effort who chose to remain anonymous. Their help is just as important as that of those above, and my heartfelt thanks is just as deep.

For my wife, Shari, and son, Alex, whose lives are intertwined with mine and, hence, with this project. The completed work is as much theirs as it is mine.

Preface

By the time I was six years old, I had lived in both a small northern Massachusetts village and London, England. The contrasts between the two are too numerous to sum up here, but the vehicle of choice while traveling to and fro was the airplane. Not the plodding, propeller aircraft, but the silver, screaming, four-engine jet that transported thousands of passengers per day over the Atlantic Ocean. My parents had separated in the late 1960s, and the jet aircraft allowed me to visit them both. Without the jet, I would have grown up not knowing one or the other of my parents.

The first crash I remember was Delta Flight 723 in Boston on July 31, 1973. All but one passenger were killed: A young air force sergeant, Leopold Chouinard, survived, but within a few months died from his injuries. Along with most of New England, I was stunned. Crashes just did not happen, I thought.

A few years later, I bought a book entitled *Disasters*. It had photocopies of the front page of the *New York Times* on the dates of infamous disasters; the *Titanic* and *Hindenburg* were prominently featured. The Delta 723 crash was also covered, as well as a crash I had been too young to remember: the 1967 midair collision of a Piedmont Airlines Boeing 727 and a Cessna 310.

The front page pictured John T. McNaughton, secretary-designate of the navy, and his wife and son. They were eating breakfast and their little boy, Theodore McNaughton, was chatting happily with his parents. A few hours later, they were dead when the jet in which they were flying struck another plane over the sleepy mountain town of Hendersonville, North Carolina, killing everyone on both planes—82 people.

After college and a stint in the United States Army, I found myself living in Spartanburg, South Carolina, just under 60 miles from the site of the crash. Talking with some locals in Hendersonville, I found the crash site and was struck by the fact that no memorial had been dedicated to the victims. In 2004, with the help of the Hendersonville community and others around the country, a memorial was dedicated two days before the 37th anniversary of the crash.

The crash happened at the height of the Vietnam War, in a summer filled with riots, at a time when Americans were questioning the motives of their own government. The summer of 1967 was also just a few months shy of some of the most traumatic experiences in American history: the Tet Offensive and the assassinations of Martin Luther King and Robert F. Kennedy. Lyndon Johnson would soon announce his retirement, and the campaign would begin that put Richard Nixon in the White House. A plane crash over a small mountain town quickly faded from memory.

As I met with family and friends of the victims who had traveled to the dedication, I was amazed at the number of questions they had regarding the crash. I too began to inquire. As I started to research the accident, I also earned my master's degree in aeronautical science from Embry-Riddle Aeronautical University while working full-time in transportation.

Most of the principal figures involved in the crash and its investigation are dead. Those who are still alive have either never responded to interview requests or have declined to speak with me.

I have relied on primary sources wherever possible, mainly transcripts of congressional hearings, National Transportation Safety Board (NTSB) hearings, the accident report, group chairman reports, court depositions, investigator's reports, and other vital government and legal documents. Secondary sources such as newspapers, textbooks, magazine articles, and interviews with next-generation family members were only used to fill in the gaps left by the primary sources or used as background.

The government transcripts of this case can be found online or in the reference records sections of many universities. They mainly encompassed testimony of those charged with aviation safety at that time: The head of the FAA, William McKee; Joseph O'Connell, the chairman of the NTSB; and various others including pilots, bureaucrats and aviation officials. Other government documents used were memos from the Federal Bureau of Investigation (FBI), the White House, the FAA, the NTSB, and the National Bureau of Standards.

The legal depositions taken in this case were uncovered in dusty boxes at the National Archives and Records Administration (NARA) in Atlanta, Georgia. Of all the depositions taken, only two remained in the public domain. The others are sealed away in the offices of those legal firms who still represent the principals involved. While a few pages of depositions outside the public realm have been revealed to me (and used within the book), the two remaining at NARA provided enough information to tell the complete story of the crash of Piedmont Flight 22.

The most important sources, however, are those of the pilots and controllers that were recorded before, during and after the accident. Not only did I retrieve the transcripts of the cockpit voice recorder (CVR) of the Piedmont jet from the NARA in Atlanta, an individual who chooses to remain anonymous sent me audio copies of the original tape recordings on disc. The conversations of the Piedmont pilots, Cessna pilot and air traffic controllers can clearly be heard on the recording, and the crash can also clearly be heard on the disc.

It is rare for a researcher to have access to so many primary sources, both audio and in print, especially after the deaths of most of those involved. It allowed me to piece together a complex and tragic accident without the prejudices and apologies that surround the use of secondary sources. This book is an effort to answer the questions of those families who lost loved ones in the crash of Piedmont Flight 22.

Introduction

This is the story of one of the deadliest midair collisions in the history of the United States and the effect it still has on American aviation policy. The collision of a Piedmont Airlines Boeing 727 and a Cessna 310 on July 19, 1967, took place over the rural mountain community of Hendersonville, North Carolina, a place affectionately known as "The Land of the Sky."

The collision came at a time of extreme upheaval and change in commercial aviation. By the mid–1960s, the first generation of American commercial jet aircraft, the Boeing 707 and the Douglas DC-8, had become huge successes with the traveling public and had established themselves as the intercontinental transports of choice. Airlines wanted to spread that success around the country and began pressuring aircraft manufacturers to build jet transport for intracontinental and regional flights. Hence the establishment of the second generation of jet aircraft, the Boeing 727, Douglas's DC-9, and the British-built BAC-111. Their role was to fly into smaller airports around the country and bring jet travel to rural America.

The Boeing 707 and Douglas DC-8 both flew in and out of the major airports around the country, mainly those in New York, Boston, Chicago, Los Angeles, etc., whose facilities were designed to handle these larger and faster jet aircraft. Smaller airports around the country were not equipped to handle them, so much so that the Boeing 727 was built with its own auxiliary power unit to supply its own power while on the ground. At these regional airports, smaller and slower propeller-driven aircraft were expected to share the sky with the faster jet aircraft, and air traffic controllers were expected to maintain control of these aircraft, many without the use of surveillance radar. This mingling of aircraft soon became a dangerous concoction, and it exploded on March 9, 1967, when a Beechcraft Bonanza propeller aircraft collided with a Douglas DC-9 over Urbana, Ohio, killing all twenty-six people aboard both aircraft.

Introduction

Throughout the spring and summer of 1967, United States aviation policy passed through one of its most intense periods of change. The war in Vietnam drained funds from aviation safety programs, air traffic controllers were overworked, airlines competed against each other for new regional routes, and President Lyndon Johnson combined all agencies responsible for transportation and transportation safety under one department. On April 1, 1967, the Department of Transportation (DOT) opened for business.

Up until that time, the Civil Aeronautics Board and its Bureau of Air Safety investigated plane crashes. Those investigative functions were transferred by President Johnson to the National Transportation Safety Board (NTSB), a small, independent agency loosely attached to the DOT for administrative and budget assistance. On July 19, 1967, the NTSB began its largest investigation yet when Piedmont Flight 22 struck a Cessna 310 over Hendersonville, North Carolina.

This is also the story of how the United States government investigates plane crashes. At the time of this particular crash, investigative techniques had been well established for decades. To this day, they remain virtually unchanged.

In its charter, the National Transportation Safety Board is charged by the United States government to investigate plane crashes and determine "a" probable cause. The government, though, had investigated plane crashes long before the creation of the NTSB. The Bureau of Air Safety (BAS) of the Civil Aeronautics Board (CAB) performed that function right up until 1967. As a matter of fact, it can be safely stated that the Bureau of Air Safety became the air accident investigative arm of the NTSB. Most of the investigative policies and procedures remained the same, and all personnel were transferred en masse from the CAB to the NTSB. Throughout the first year of the NTSB's existence, witness statements were still being taken on forms with a CAB heading.

The primary method used by the NTSB to investigate plane crashes is called the "party system." An extremely controversial method, it allows those entities involved in the accident to help investigate the accident. Put another way, those who stand to lose most from possible adverse findings against their company are allowed to participate in the investigation, while survivors or family members of those lost are not.

If a Boeing aircraft crashes, then personnel from Boeing are picked by the NTSB to participate in the investigation. If that aircraft was equipped with Pratt & Whitney engines, then Pratt & Whitney would also be a part of the investigation. The same goes for the airline involved, and the Federal

Aviation Administration is always a part of the investigation. Each of these people is assigned by the NTSB's lead investigator to an investigative group. Groups are generally broken down to different areas involved in a plane crash. There is an Air Traffic Control group that investigates the actions of the controllers and a systems group that will analyze the different systems on the plane. Also maintenance, human factors, witness, and weather groups are formed, along with other groups, to investigate the different aspects of the flight. There is no magic formula to balance a group equally among party members so that one entity may not have more representation than another. One party may grossly outweigh another in assignments. That is the way the NTSB always investigated accidents, and it remains so to this day.

This is also the story of blame, or rather, the search for blame. The financial stakes are huge when planes crash. Those involved in a crash will do everything they can to argue that the accident was not their fault. That is understandable: it is human nature. In addition to economic stakes are political stakes. Make no mistake, plane crashes can quickly become political issues. Never was it more so than with the Flight 22 midair collision.

By 1967, it was clear to everyone that government funds were being diverted from domestic programs to fund the Vietnam War, and no one knew that more clearly than members of the House of Representatives Aviation subcommittee; the chairman of the FAA, William McKee; and President Johnson. Within twenty-four hours of the Piedmont crash, congressional leaders called for an emergency session to discuss aviation safety. With the government in charge of the safety and operation of the nation's airways, funding, or lack of it, would be a vital topic in those hearings. The leaders of American aviation trekked to Capitol Hill and testified before Congress. Their testimony revealed an aviation system antiquated by lack of modern equipment, underfunded by a drain of government money to Vietnam, understaffed by an overworked group of dedicated air traffic controllers, and unappreciated by a president trying to give his people "guns and butter," and realizing, almost too late, that there was not money for either.

Plane crashes happen quickly, and their appearances as news stories just as quickly disappear as other news stories replace them on an hourly cycle. That is what happened with this crash. War, assassinations, elections all led to the erasure of this crash from the public's memory. However, the inconsistencies of the NTSB's investigation, the damning testimony of aviation officials on Capitol Hill, and the courage of a few public servants all combined to force me to take another look at the crash and how the United States funds its aviation system and investigates its aviation accidents.

1

"I HAVE A FLIGHT PLAN FOR YOU"

Bulging beneath the belly of the Boeing 727, the tires of the descending aircraft caught the ground with a surprised gasp, as if startled by the runway's sudden appearance. The flexing curved wings strained against the flow of air, forcing the jet to slow down as this leg of the trip ended. Within seconds, the plane had slowed enough that maneuverability was nothing more difficult than driving a car in a parking lot. At 11:42 a.m., July 19, 1967, approximately twenty minutes behind schedule, tower personnel at the Asheville, North Carolina, airport cleared Piedmont Flight 22 to proceed to the ramp, ending its first hop of a three-stop flight from Atlanta to Washington, D.C.

"Do all that rushing and still lose it in the traffic pattern,"[1] First Officer Thomas Conrad complained aboard the 727.

"Well, we made our estimate by—I missed it by one minute, so that's not so bad as hell,"[2] Captain Ray Schulte answered as he maneuvered the plane to the ramp.

Two minutes later, airport attendants kicked a pair of chocks onto the wheels of the aircraft, and the crew of Piedmont Flight 22 killed the engines and waited as the two flight attendants in the back directed thirty-six exiting passengers off the aircraft. As quickly as the departing passengers were off, twenty-three new ones got on. The exchange had taken just a few minutes, but the flight was still running behind.

Airline passengers appreciated on-time flights. Airline management appreciated pilots who delivered their passengers on time. Most airlines, including Piedmont, collected and maintained information regarding schedule delays.[3] Flight delays at Piedmont were divided into two categories: recurring delays and incurred delays. Recurring delays were evaluated to try to identify their cause. If a given pilot was continually delayed, action would be taken. Incurred delays are mechanical or servicing delays. Either way, even though airline management dismissed the notion that they applied pressure on pilots to maintain their schedules, pilots did not want their flight numbers attached to either list.[4]

Captain Schulte and First Officer Conrad knew exactly how important this flight number was to Piedmont Airlines. This was the newest jet route for Piedmont from Atlanta into Washington, D.C. This was also the first jet aircraft maintained by Piedmont. This was a high-priority and extremely prestigious run, as it connected several southern cities to the nation's capital. It also connected Washington to the South, and at any given time there could be one, or several, public figures on board. By providing jet service to and from Washington, Piedmont was banking that it would be the choice of passengers flying that route.

The seventy-four passengers now sitting inside the 727 were divided by a single aisle down the middle of the aircraft. Three passengers could sit in each row on the left side of the aircraft; two passengers could sit in each row on the right side. The jet's capacity was ninety-four passengers.

This particular jet, registration number N68650, had just turned four years old. In its short existence, N68650 had led a cosmopolitan lifestyle. After its birth in Seattle, the aircraft was leased by Iran National Airlines, and joined the fleet of the Shah. After a few years jetting around the Middle East, N68650 returned to its birthplace and waited for someone to adopt it. A short while later, the management of North Carolina-based Piedmont Airlines had decided that the Boeing 737 aircraft would best suit their business needs, but none were available. As part of a deal, Boeing leased N68650 to Piedmont to tide them over until the 737 became available. The adoption was finalized on February 25, 1967.[5]

After a Winston-Salem christening, N68650 received a new name. In keeping with Piedmont's tradition of naming an aircraft a "Pacemaker," they named it *Manhattan Pacemaker*, as it would eventually connect New York City with the southern United States.

The *Manhattan Pacemaker* cut an impressive figure in the southern airports it serviced. From the ground to the tip of its tail, it stood 34 feet tall. At over 133 feet in length, it beat out the largest propeller planes by at least 15 feet. With its "T" tail configuration, it possessed an incredibly high rate of descent and climb, and with the three Pratt & Whitney JT8D engines sandwiched around its tail, N68650 had a maximum takeoff weight of 170,000 lbs.[6] As compared with the largest of the piston-driven planes currently in service, the 727 was only a little bit bigger, a little bit taller, but

1. "I Have a Flight Plan for You" 9

a whole lot faster. In fact, its speed became part of its popularity. It could soar 200 miles per hour faster than the DC-7 or the L-1049 Super Constellation, the newest of the prop planes. The 727 could take more people, more quickly, to any part of the country.

The race to acquire jet aircraft had started approximately ten years earlier with the introduction of the Boeing 707 and the Douglas DC-8. Those planes were too large for the routes Piedmont Airlines traveled, but the 727 was perfectly suited for the small airports around the South, the area Piedmont prided itself on servicing. The reason for the 727's suitability came from its innovative flap system, which increased the area of its wings by 25 percent during takeoff and landing. This gave the aircraft the ability to utilize airports where accessibility was limited, which became a huge advantage at the smaller Appalachian mountain airports Piedmont serviced. The 727's independent auxiliary power unit meant it could land at airports that did not have the capability to power a craft on the ground.

While the innovative flap feature on its wings opened up new destinations to Piedmont, it did not come without growing pains. Just over two

Publicity photograph of the *Manhattan Pacemaker* in Piedmont Airlines livery.

years after its introduction, four 727s crashed on approach to different airports around the world. Pilots did not anticipate the quick rate of descent inherent in the 727's design, and they found themselves on the ground before they realized it. Some members of Congress called upon the airlines to withdraw the aircraft from service, but in response the airlines stepped-up their training programs, placing an emphasis on safety and cockpit awareness.

While technological advances in aircraft increased, the airports they serviced tried to keep up. Some simply were not ready for the new aircraft, but since the Federal Aviation Administration (FAA) promoted commercial air travel, it did nothing to restrain the introduction of these 550-mph aircraft into the sky alongside the slower piston-driven ones, even into airports not protected by radar.

Soon after its arrival at Piedmont's headquarters in Winston-Salem, North Carolina, the *Manhattan Pacemaker's* charm quickly faded. On May 25, 1967, as it was being towed from the Smith-Reynolds Airport in Winston-Salem to a hangar across the street, a car struck the underbelly of the aircraft and careened down its fuselage, causing $25,000 in damage.[7] As recently as July 14, 1967, as N68650 approached Smith-Reynolds airport in Winston-Salem to land, a warning light pierced the relaxed cockpit environment, notifying the crew that the nose wheel had not come down. After a spin around the airport, tower personnel notified the crew that they could observe the wheel and it had locked in place for a safe landing. However, the crew had already taken the precaution of explaining to the passengers how to brace themselves in the event of a hard landing. All the assembled personnel on the ground, fire trucks, police, ambulances, and other airport personnel breathed a sigh of relief as the plane landed smoothly. After a quick inspection by the mechanics, the aircraft flew on to Asheville, where, again, the warning light came on. The crew immediately turned around and landed back in Winston-Salem in front of the same audience as it had a few hours before.[8] Mechanics identified the problem and fixed it, but already it seemed like the aircraft was unlucky. Piedmont pilots certainly hoped that the *Manhattan Pacemaker's* luck would hold until the new 737's arrived the following March.

The crew that had just landed the plane in Asheville on July 19 had been together since early that morning, having started their workday at National Airport in Washington, D.C. Captain Raymond Schulte, age 49, led the group. Schulte began his flying career on November 14, 1940, receiving his training from the civilian pilot training courses of that period. He acquired a commercial pilot's certificate in 1942 and began working for

Piedmont Airlines on July 18, 1947—literally a founding pilot of the airline. As Piedmont grew, Schulte grew with it, receiving pilot-type ratings on the DC-3, Fairchild F-27, Martin 202/404, and the Boeing 727.[9] The transition from piston aircraft to jet aircraft did not go without a hitch for Captain Schulte. He completed his 727 training on May 10, 1967. He received an unsatisfactory grade on his initial rating check. He had problems in the traffic control and holding areas. "He repeated the maneuver on a re-check the following day and received a satisfactory grade."[10] He passed his last line check on the 727 in early July and had only 151 hours of flying time in that aircraft as compared to his total flight time of 18,383 hours.[11] Among Piedmont personnel, Schulte was known as "a competent and conscientious pilot who was very fair in dividing piloting time equally with his copilot."[12] Piedmont Airlines hired First Officer Thomas Conrad on March 30, 1961. He had 3,364 hours of total flying time. He began training on the 727 on April 11, 1967, and completed the course ten days later. He passed his First Officer's check with above average grades. His total flight time in the 727 came to just 135 hours.[13] Employees at Piedmont regarded Conrad as a "likeable, witty extrovert who was a dependable co-pilot."[14]

Ironically, Second Officer (Flight Engineer) Lawrence Wilson had claimed just about as many hours in the 727 as Conrad and Schulte combined. He had racked up 281 hours in the three-engine jet. Piedmont had hired Wilson less than two years earlier, in August 1965. According to investigative records, "He attended the Boeing Factory School in the fall of 1966 prior to acquiring his Flight Engineer's certificate."[15] He claimed 7,754 hours of total flight time. Among Piedmont crews, Wilson earned a reputation as a highly competent flight engineer, someone who knew his job well and on his last proficiency check scored "well above average in the emergency and abnormal operations area."[16] Sandra "Kay" Cox and Deborah Davis, 22 and 20 years of age respectively, drew flight attendant duties for Flight 22. Each had just over six hours of flight time on the 727 and both had only been working for Piedmont for a couple of months.[17] "God almighty, seventy-three souls,"[18] Captain Schulte exclaimed after he received the dispatch release and passenger count. He had received incorrect information. The count actually came to seventy-four passengers.

"What day is today? Tuesday, ain't it?"[19] Schulte asked.

"Wednesday, Wednesday,"[20] Conrad responded as Wilson checked the gauges in front of him.

"Came out here with seventy,"[21] Schulte responded, surprised at the high passenger count.

Western North Carolina experienced a surge in visitors during summer months every year. The opening of summer camps all around the mountainous region attracted thousands of campers. Piedmont and other airlines serving the region added additional passenger service to the area every summer just to shuttle campers to and from the area. Asheville had also started gaining popularity as an inviting place to hold conventions. The National League of Insured Savings and Loan Associations had just wrapped up its national convention in Asheville that morning. It seemed Asheville gained in prominence every year as a great area for both business and pleasure.

Schulte glanced at the dispatch papers. Visibility notched at four miles, with haze, a dew point of 63, wind at 160 degrees at five knots, rain zero with a temperature of 74 degrees. Broken clouds hovered at 2500 feet.[22] Liking what he read, he gave the command to start the engines at 11:50 a.m.[23]

Approximately thirty minutes earlier, David Addison had started the engines on his plane at the airport in Charlotte, North Carolina, approximately 120 miles to the east of Asheville. His two traveling companions, Robert Anderson and Ralph Reynolds, waited patiently as Addison began final preparations. After less than twenty-four hours in Charlotte, North Carolina, Robert Anderson, 36 years old, could not believe a region could claim the countenance of hospitality yet still segregate human beings because of their color. He was known as an avid whistler, but *Dixie* would not claim a position on Anderson's Hot 100 list.

Their southern journey began two days before on July 17, 1967. Lanseair, a Springfield, Missouri, company specializing in underwriting transportation insurance companies, wanted to introduce Anderson as their new safety engineer to its clients in the South. Anderson was teamed with Ralph Reynolds, Lanseair's claims department manager, and the two, along with their hired pilot, David Addison, 48, left Missouri to meet with customers in North and South Carolina. Anderson's disgust at southern hypocrisy started as he observed a court session in Charlotte to which Lanseair was a party. Anderson looked on with embarrassment as blacks sat in the back of the courtroom while white folks crowded the front. He spoke to his wife later that evening, recounting to her the segregated scene.[24]

With flying quickly becoming the mode of choice for business travel, Dave Addison found himself frequently employed by Lanseair as a pilot-for-hire. An employee of Rapid Air, he had flown many employees of Lanseair and knew the owner, Joe Dando, on a first-name basis. No novice to flying, Addison received his pilot's license in 1941. In 1942, he earned his commercial pilot's certificate and gained his flight instructor rating a

1. "I Have a Flight Plan for You" 13

few months later. After the war, he earned his ground school instructor rating, which allowed him to teach pilots in the areas of aircraft, aircraft engine, civil air regulations, meteorology, navigation, and radio navigation. Within a few years, he received an advanced ground school instructor certificate. David Addison knew how to fly planes, having accumulated over 10,000 hours as a pilot-in-command by July 19, 1967.[25]

That morning, Addison requested and received two weather briefings via telephone. Addison, Anderson, and Reynolds had earlier discussed their destination for that day. Anderson wanted to fly directly home to Springfield that day, but Reynolds demurred and requested that they land at Asheville to introduce Anderson to some clients. It would be a quick stop and probably to their benefit, because an earlier weather briefing indicated that the main weather problem along their route was "stratus and fog over the western portion of the Carolinas and eastern Tennessee."[26] A stop in Asheville would give the fog time to dissipate, and they could continue onto Springfield later in the afternoon. The three men decided to stop in Asheville.

Many companies, like Lanseair, picked the Cessna 310 as their plane of choice because of its speed and capability. Their Cessna 310, N3121S, shined brightly on the runways and in the sky. Bright red paint covered its exterior, with cream-colored stripes unfolding along its sides. The 310 could reach speeds of 245 miles per hour and held up to six people comfortably. Car trips that took a day turned into hour-long jaunts, which saved money on hotels, fuel, and meals. It also made for happy employees. They could be home more nights with their families.

At 11:22 a.m., Addison asked for another weather report while he steered the plane onto the taxiway at the Charlotte airport. He did not file a flight plan to Asheville, nor was one needed. Supervisory Air Traffic Control Specialist Benjamin Folk worked the local control position in the Charlotte tower and answered Addison. Folk reported broken clouds at 2500 feet in Charlotte and scattered clouds in Asheville at 2000 feet, allowing pilots to operate under visual flight rules (VFR).[27] This concept put the onus for traffic control onto the pilot and crew. They inherited the responsibility to look out for other aircraft, a concept called "see and avoid" or "see and be seen."

Addison's experience never left anything to chance while flying. To get through the thin layer of clouds, he requested an instrument flight rules (IFR) clearance "to on-top." There is no indication anywhere that Addison wanted to fly IFR to Asheville as opposed to visual flight rules (VFR).[28] Flying IFR meant relying on cockpit instruments rather than sight during

A 1955 red and white Cessna, almost identical to the accident aircraft, 3121S (courtesy Carl Evans).

flight and filing a flight plan. Addison knew he did not need a flight plan to travel through the thin layer of clouds. He merely wanted to get on top of the clouds and then fly VFR to Asheville.

The request confused Folk. His responsibility that day was for local control of aircraft at the airport. Ronald Painter, air traffic control specialist, asked Folk to cover his duties for a few minutes. So at the moment Addison requested "IFR to on-top," Folk was now performing two duties at the airport—local control and ground control. When Painter returned, Folk relayed Addison's request to Painter, who took over ground control.[29]

Painter informed Addison he would have to file a complete flight plan to fly "IFR on-top."

Addison mildly protested. "Do we need a flight plan for just—ah—IFR to on-top?" "To get IFR on-top, I'll need a complete flight plan on you," Painter replied.[30]

After his feeble protest, Addison complied with the controller's command and filed a complete flight plan requesting 8,000 feet, Victor 296 to Asheville. A pilot with over 10,000 hours, Addison gave the controller every item required for an IFR flight plan.[31]

At 11:29 a.m., Painter cleared Addison "to the Asheville VOR, direct

1. *"I Have a Flight Plan for You"*

route, maintain six thousand, turn left on course."[32] At 11:30 a.m., Addison coaxed the Cessna down the runway, and less than a minute later the aircraft lifted off the ground and eased itself comfortably into the sky.

All flights around the southeastern United States were coordinated through the Atlanta Air Route Traffic Control Center in Hampton, Georgia. Their radar monitored all flights in the region, but there were exceptions due to the vastness of the area. Most of these gaps were filled by terminal radar units within the vicinity of airports. If an airport had its own terminal radar unit, its own personnel could monitor incoming and outgoing flights without help from Atlanta. The Asheville airport, though, did not have radar, and Atlanta radar did not cover parts of the area Addison to which was heading.

"I have a flight plan for you, Atlanta," a voice from the Charlotte sector phoned in.

"Whom am I talking to?" Verl Hawkins replied from the Atlanta Control Center.

"Charlotte."

"Go ahead."

Hawkins listened intently as Addison's flight plan was relayed to him. "3121 Sierra off Charlotte at thirty," the voice told him.

"OK," Hawkins replied as he prepared to receive the aircraft on his radar screen. The blip popped up on his screen within five minutes, and Verl Hawkins had control of the Cessna now.[33]

For the next several minutes, Addison listened and executed every command that Hawkins issued.[34] The Cessna cruised towards Asheville at approximately 200 miles per hour. "Cessna 21 Sierra is cleared to the Asheville VOR," he told Addison. "Descend and maintain seven thousand. Expect ILS approach at Asheville."[35]

"Leaving eight at this time," Addison replied.[36] He extracted the ILS approach plate for Asheville from its case and studied it for a moment. This would be his bible for a safe landing in Asheville. He could check his minimum altitudes and any other pertinent information to effectively execute his ILS approach and landing. Nothing on it confused Addison. He had looked at thousands of approach plates before.

Less than a minute later, Hawkins cautioned Addison. "Cessna two one sierra, your transmissions are hard to read. You say you are leaving eight thousand now?"

"I left eight thousand for seven," Addison replied, making sure Hawkins knew he had already started his descent.

"The traffic for two one sierra is now at your twelve o'clock position about three miles northwest bound," Hawkins replied, warning Addison that there was another aircraft within his airspace.

Addison scanned the horizon. "Negative contact. We're in the clouds." He turned the dials on his transmitter in the cockpit, trying to fine-tune his communications with Atlanta. "Atlanta Center. Two one Sierra. Is this transmitter any better?"

"Two one sierra. It's a little bit better," Hawkins replied.[37] As Hawkins kept the two aircraft separated so that no collision would occur, he marveled at the technology of the radar. He and many others in the air traffic control field quickly realized the importance of such a wonderful tool. He hoped the time would soon come when all airports were equipped with the device.

"Radar service terminated now. Contact Asheville approach one two five point three," Hawkins instructed, handing Addison off to Asheville.

"[Unintelligible] point three," Addison replied, repeating the radio transmitter code.[38] He relaxed as the flight continued. Except for the controller in Charlotte, it seemed from his professional point of view that the traffic controllers knew what they were doing. With the notification of what type of approach to expect in Asheville, the rest of the flight should be uneventful, he thought. As uneventful as the thousands of others he had flown before.

2

Passenger List

The hum of excitement grew louder as takeoff time approached. Those sealed within the 727 cabin chatted with each other like long-lost friends. In fact, many of them were friends. A group of men, making up over one-third of all the passengers, talked about the good times to be had when they arrived at the Greenbrier Hotel in White Sulphur Springs, West Virginia, for the annual Stokely food brokers convention. This group, the last arrivals for the convention, represented the southern and western United States, from Florida to New Mexico.

James Chidsey co-owned Chidsey & Schroeder, a small warehouse business that carried name brands of southern products like Dixie Crystal and Stokely-Van Camp. When orders for various products arrived at the warehouse, he would personally load the items in his van and deliver them to stores in Tennessee, Georgia, and North Carolina.

Chidsey knew the Blue Ridge Mountains. He admired the view as the jet landed in Asheville from Atlanta. It contrasted sharply with the Colorado Rockies, from which he had just returned from vacation at a dude ranch with his wife and kids. The ranch may have been an unconventional place to spend vacation, but the time together did the family wonders. They had already started talking about another family vacation later in the year, perhaps in the fall.[1]

Webster Benham moved to Albuquerque, New Mexico, from Oklahoma in 1951 and became a senior partner in the food brokerage firm of Benham-Bryant. Just seven years prior to that, he represented Oklahoma as its Outstanding Young Man of the Year. During the fifties, Benham became heavily involved in the Albuquerque civic scene. When the city needed a cash infusion to stimulate industrial development, Benham led the effort, becoming the first to pledge one thousand dollars. He also solicited large contributions from other brokers in the area, heading up a team called "Benham's Bruisers." As a reward for such a successful fundraising effort, city officials named Benham president of the Executive Association of Greater Albuquerque. His term expired the day he left for the Greenbrier.[2]

Two-thirds of the passengers were not food brokers. Some, like fourteen-year-old Martin Shuler and his grandmother, Beulah Lance, boarded the plane to explore the historic sites in Washington, D.C. Shuler had a fascination with history, and his grandmother offered to chaperone his trip. Flying there on a jet airplane only made the trip more exciting for the young boy.[3]

What was left of the Love family of Jackson, Mississippi, had boarded the 727 in Atlanta. Just two months prior, Dr. William Love, a heart surgeon, died in his classroom at the University Medical Center in Jackson of a massive heart attack. Mrs. Love wanted to visit her husband's parents in Richmond and grieve with them over their shared loss. With her in the cabin sat her daughter Lucy, age 19, daughter Ellie, 16, and son William Love V, age 6. One other daughter had left for vacation earlier to visit her other grandmother in Minnesota.

While one-third of the cabin talked about an upcoming convention, the other two-thirds might well have been discussing war in Vietnam. Indeed, almost every American would hear it mentioned during the course of any given day, as each day brought more ominous developments with America's ever-deepening commitment to its South Vietnamese ally, and an ever-growing body count. Antiwar protests grew larger and louder daily. Celebrities and politicians called for the end of the war. The war dug deeper into the American psyche every day, and Pentagon officials knew privately what they were afraid to say publicly: that the war strategy was not working and prospects for victory grew dimmer by the hour. Some in the Pentagon privately told President Johnson that the war was not winnable, but a major breach within the administration had not occurred yet.

One man sitting next to his wife and son onboard Piedmont Flight 22 knew that the breach had slowly begun to open and within a few months would burst open like a cracked dam. As the closest civilian aide to Secretary of Defense Robert McNamara, recently confirmed secretary of the United States Navy John T. McNaughton had been asked by McNamara to widen the breach.

Born in Pekin, Illinois, the son of the town's largest newspaper owner, McNaughton graduated from Harvard, fought in World War II, got married, ran unsuccessfully for Congress, and returned to Harvard to teach, all before he turned thirty-five. In 1961, as John F. Kennedy's New Frontier began, Secretary of Defense Robert McNamara picked McNaughton to come to Washington and become his assistant secretary for international affairs. His role in that office led to arms reduction talks with the Soviets and to the establishment of the "hotline" between Washington and Moscow. But all of his accomplishments, and the accomplishments of the Kennedy and Johnson administrations, both foreign and domestic, were coming undone due to the unpopularity of the Vietnam War.

McNaughton sensed McNamara's frustration and probably calculated that the defense secretary would soon leave the Pentagon. He did not know whether that removal would be voluntary or involuntary. As the number of dead American soldiers increased steadily to a consistent one hundred per week, McNaughton knew that fingers would be pointed and officials would duck for cover or shift the blame to others. Smart enough to light both ends of the candle, McNaughton's public position and private position on the war were dramatically opposed. Specifically, he had advised that air strikes on North Vietnam might be a way to help bring an end to the war, but he privately ridiculed the idea even as he planned them.[4] By this point in time, McNaughton knew that the war was lost, and thinking of his future political plans, he began thinking of a way to distance himself from it.

Then, incredibly, Paul Nitze resigned his post as secretary of the navy. McNaughton, trained in the art of the possible, quickly realized the implications of inheriting that position. The post was mostly ceremonial, but he could still stay on as an advisor at the Pentagon. He could urge McNamara and Johnson to end the war without leaving any fingerprints. The Senate confirmed McNaughton as navy secretary, and after that grueling process, he joined his wife, Sarah, to pick up their son Theodore from a summer camp in the mountains of western North Carolina. The trip afforded a needed rest and a quiet place to think.

McNamara's request to begin a study on Vietnam, to begin widening the breach, bothered McNaughton. As a lawyer, he knew panic when it showed, and it had shown up in McNamara. McNaughton knew McNamara wanted documents, documents dating back to Harry Truman's time in the White House all the way through to the present. McNamara wanted these documents so that future scholars could understand and study what went wrong in Vietnam. McNaughton had been around long enough to know that Johnson would blame McNamara for the quagmire in Vietnam, and McNamara wanted the documents to prove the president wrong. McNamara's order still sent chills down McNaughton's spine.[5]

McNamara did not want to be involved with the project. "I told him to cast his net wide," McNamara related years later. "Let the chips fall where they may."[6]

McNaughton did not ask if the president knew about it. Since McNamara did not say, McNaughton did not ask. The documents could be severely embarrassing to Johnson or, worse, drive him from office. John McNaughton knew that no one, and especially not him, would tell the president about this undertaking.

Was the order to prepare the study treason? No. Could it be considered disloyal to the president? Sure. McNaughton knew he had to move carefully. Three weeks before, he had outlined his plan to collect the documents and knew it had to be top secret. If they fell into the wrong hands, the documents could be devastating because they were full of lies and deceptions dating back over twenty years. An evidence scholar, McNaughton knew how to keep secrets and knew how to keep files secure.

Staring out the thick Plexiglas window of the 727, McNaughton saw the Blue Ridge mountains to the west of the airport. Densely forested, they resembled the mountains he had seen on his fact-finding trips to Vietnam. Even here, listening to his son chatter about his summer camp activities, he could not escape thoughts of Vietnam. And now he was flying back to the one place where he could not escape Vietnam at all.

3

"TWENTY-TWO IS READY TO GO"

If Margaret Mitchell had written a book about the southern way of life in the twentieth century, certainly a mention of Piedmont Airlines would be found in it. Born in 1948 among the craggy tobacco fields of eastern North Carolina, Piedmont cultivated its image as a southern institution.

Starting on February 20, 1948, Piedmont Airlines Flight 41 took off from Wilmington, North Carolina, en route to Cincinnati, Ohio, connecting the Ohio River Valley with the Tidewater area of North Carolina and Virginia. The DC-3 made several stops that day on its way to Ohio. The stops were not at large airports but rather places like TriCities Airport in Johnson City, Tennessee, and A&H Airport in Asheville. These smaller airports fed the larger airports passengers, who in turn went on to more distant destinations. Hence Piedmont became known as a feeder airline. By 1955, the airline had grown to eight hundred employees and a fleet of 16 DC-3s.

H. K. Zeke Saunders, vice president of Piedmont Aviation and former Piedmont chief pilot, attributed part of the early success of the airline to the pilots who were hired. "We were very particular about who we hired as pilots. We were looking for a particular kind of personality. Calm ... and from a certain kind of background: stable. We wanted calm people who came from stable families."[1] So why Piedmont assigned Captain George Lavrinc, thirty-two years of age, as pilot in command of Flight 349 from Washington, D.C., to Roanoke, Virginia, on October 30, 1959, is anyone's guess.

Piedmont hired George Lavrinc in 1951. In 1957, he received a promotion to captain. On the day before Halloween 1959, Lavrinc found himself in command of Flight 349. The flight was scheduled from Washington, D.C., to Roanoke with a stop in Charlottesville, Virginia. While attempting to land in Charlottesville, Lavrinc slammed the plane into Bucks Elbow Mountain, thirteen miles west of the airport. All but one of the twenty-seven people aboard, including Lavrinc, died.

After an intensive investigation by the Civil Aeronautics Board (CAB), investigators determined that the cause of the accident was a "navigational omission resulting in a lateral course error that was not detected and corrected through precision instrument flying procedures."[2] In other words, Lavrinc turned about eight miles away from where he should have. Why the pilot did this puzzled investigators until they pried into his background. There they found Lavrinc's life rocked by instability consisting of a crumbling marriage, religious zealotry, and prescription drugs.

By the end of 1957, Lavrinc had filed for divorce. When he transferred to Washington, he left his family in Norfolk. He became a member of the fundamentalist Cherrydale Independent Baptist Church in Arlington, Virginia. He devoted much of his spare time to his religion, working with youths and alcoholics. He also helped conduct services at the local jail, joined the Youth for Christ Movement, and tried to sway others to his religious beliefs.[3]

In the fall of 1958, he enrolled in Washington Bible College, and the added burden of studies, on top of his divorce and his job as a pilot, increased his stress. He complained of headaches and had an operation for a sinus condition. He wanted to reconcile with his wife, but only if she could join him in his religious beliefs. He even asked Piedmont airlines for a two-year leave of absence so that he could become a flying missionary in South America.[4] Then, as suddenly as he found religion, his zeal for it waned. He consulted psychiatrists, and they prescribed him tranquilizers. The CAB determined that Lavrinc was "so burdened with mental and emotional problems that he should have been relieved of the strain of flight duty while undergoing treatment for his condition."[5] The board also made a simple determination at the end of its official report: "If a crewmember's personal situation demands tranquilizers, he should be removed from flying status while on the drugs."[6] Put simply, Piedmont Aviation and its officers should never have allowed Captain Lavrinc to fly after August 1959, the first time the psychiatrists prescribed his drugs.

The crash of Flight 349 quickly receded into the past. The aging DC-3 fleet needed to be replaced, and in 1956, Piedmont's founder, Tom Davis, decided on the F-27 aircraft. These aircraft were air-conditioned, a valuable commodity in the South during the summer. The aircraft also had a 36-

seat capacity, an increase of fifteen per flight from the DC-3, and could travel 280 miles per hour, an increase of 50 miles per hour from the DC-3. As well as a pressurized cabin, the aircraft came equipped with the all-important weather radar so that it could avoid the sudden and deadly storms that swirled above the mountains along Piedmont's routes.[7]

In 1961, Trans World Airlines placed its fleet of Martin 404 aircraft up for sale. This propeller aircraft could seat 44 passengers and traveled 100 miles per hour faster than the DC-3. Piedmont bought 17 of these aircraft at a cost of $6 million, an incredible bargain.[8]

As the middle of the decade arrived, a decision loomed for all airline bosses: what jet aircraft would they choose? Would it be the British BAC-111, an eighty-five-seat design? Or the Douglas DC-9, a twin-jet American-built plane seating 90 passengers? Or the Boeing 737, a twin-engine seating ninety-five? Each airline boss considered the special requirements of his respective airline, and each chose according to his needs.

In June 1966, Tom Davis announced that Piedmont Airlines would purchase the Boeing 737. Davis based his choice on the merits of the aircraft's newness, durability, ruggedness, and great field performance.[9] Davis, however, would have to wait, because the aircraft would not be available until early 1968. But in the latter part of 1966, the CAB authorized Piedmont to serve New York City; Davis decided he could not wait to fly jet aircraft any longer, especially into the Big Apple. So he authorized the lease of two Boeing 727 aircraft, and on March 15, 1967, Piedmont entered the jet age with N68650, the *Manhattan Pacemaker*, an aircraft that had a speed more than double that of the DC-3.[10] In comparing the 727 to the DC-3, Captain Leon Fox, the pilot of Piedmont's first commercial flight back in 1948, stated, "Brother, that was a lulu. It was the speed. You had to keep thinking all the time to stay ahead of that airplane. You had to stay ahead of it.... In the DC-3 you could just kind of dive down and flare out at the bottom and land it, but a 727 won't flare out; it'll keep diving right into the ground."[11]

Air traffic controllers' responsibilities in 1967 had increased exponentially with the explosion of commercial air traffic. Juggling incoming aircraft while at the same time manipulating flight paths in coordination with Atlanta air traffic control, all the while without terminal radar, could suck the lifeblood out of traffic control specialists like James Watkins. The multiple tasks demanded of a controller seemed unending, and with the bitter fight over staffing and funds between controllers and the Federal Aviation Administration at a boiling point, the future did not look bright. Low morale brought a decrease in the safety of the skies. James Watkins could not allow that, and so he performed his duties without complaint. Besides, he knew a change was coming for him with his recently approved transfer down to the airport in Greenville, South Carolina.[12] Even if the workload remained the same, a new job at a new location offered a refreshing change in a burn-out career.

"Do you know where this Cessna two-one Sugar is right now?" Watkins asked Atlanta control on the telephone.

Before Atlanta could respond, Watkins heard Addison on the radio giving his position. "Disregard, he's coming."

"3121 Sugar. What radial are you passing now off Spartanburg?" Watkins asked.

"Stand by," Addison replied as he checked his location off the Spartanburg VOR. "We're on 340," he added as soon as he found it.

"OK," Watkins answered.[13]

Watkins had help today. Evan Brown and Jesse Welch were both in the tower with him. Brown pulled the local control and ground control responsibilities, while Welch worked the en route position. The team had worked together before, and they had faith in each other's abilities.

Watkins' position would be the busiest, as he coordinated the incoming planes on the radio while at the same time keeping Atlanta control apprised of the traffic situation. Though airport traffic increased this time of year due to the influx of campers who infiltrated the Asheville mountains, Atlanta seemed to be on top of everything today and had ready replies to Watkins' queries.

Atlanta rang again. "Clear Piedmont Flight 22 slant alpha to Roanoke Airport, via direct Valdese, J-53, flight plan route, maintain 210,"[14] the caller advised Watkins.

Watkins turned to relay the information to Evan Brown, but Brown excused himself to go to the bathroom downstairs. Welch took over Brown's spot, doubling his duties for the next few minutes. Welch relayed the information to Flight 22.

"Piedmont 22; Roanoke Airport direct Valdese J-53; flight plan maintain 210."[15] Schulte repeated the instructions, obviously liking them because they gave him considerable leeway to get to Roanoke.

No one could call the instructions given to Schulte a clearance. They did not clear him for anything. As a matter of fact, they allowed him to fly to the Valdese Intersection virtually unrestricted. As Roys C. Jones, director of the Air Traffic Control Department for the Aircraft Owner and Pilot's Association (AOPA), wrote less than a month later, "Neither the pilot, nor the tower, nor the Atlanta Center controller expressed any idea, loose or precise, as to how the pilot is going to navigate directly to a VOR Intersection, Valdese."[16]

Any aircraft flying where it should not be within this area would be threatened by the freedom the tower just bestowed on Flight 22. Jones noted that the Valdese intersection itself is more than fifty miles from the airport in the complete opposite direction from the 727's takeoff. According to his instructions, Schulte would basically have to use dead reckoning to get to the intersection "direct."[17] No jet pilot was that good.

Meanwhile, south of the Asheville airport, Captain Paul Snell of Piedmont Airlines steered his aircraft, a Martin 404, towards the airport at a height of seven thousand feet. Piedmont had assigned Snell's aircraft the flight number 1022, and it carried a group of summer campers to the western North Carolina mountains.

Watkins needed to clear Snell for landing on the runway because it was the same runway that Schulte needed to take off on. "Piedmont 1022, descend to 600. Cleared for an ILS approach; plan to circle runway 16."

Snell acknowledged. "1022."[18]

Within the cockpit of Flight 22, the pilots argued mildly between themselves. "J. C. baby, who's driving?" Captain Schulte asked. "[It's] my leg to sleep, dammit."

"Shoot, I could use some myself,"[19] Conrad answered, acknowledging his own fatigue as they slowly started to taxi from the gate. Conrad would be flying the plane to Washington.

Lawrence Wilson got in on the conversation. He turned to Schulte and Conrad and asked them their dinner plans for this evening. They all decided on steak but did not know whether they wanted to go out to eat or cook it themselves.[20]

The banter subsided as the crew finished up their preflight responsibilities. Satisfied they had done everything to ensure a smooth and safe flight, Conrad called the tower. "Twenty-two is ready to go," he said firmly.

"Piedmont 22, hold," Jesse Welch responded from the tower.

"Aw shit!" replied Conrad from the cockpit.[21] The crew of Flight 22 did not want to be any later than they had to be. They had to make up some time from the delay they incurred on the flight from Atlanta.

With so many "22s" being thrown around between the tower and the 727, Captain Snell became confused. "Are you calling Piedmont 1022?" he asked the tower.[22]

Still performing en route and local duties within the tower, even Jesse Welch became confused as to whose voice came over the radio. "Piedmont twenty-two, Piedmont twenty, ten-twenty-two. Are you on this frequency now?"

"Roger," Snell replied.

"Say your altitude."

"Six thousand," Snell responded from his Martin 404, still about ten miles away from the airport.

"OK. Circle to land runway one six. Wind one eight zero degrees at eight and say position now," Welch commanded.[23]

Schulte, Conrad, and Wilson could hear the conversation between Snell and Welch on the company radio located within the cockpit. They, however, were engaged in another checklist. They shared portions of the list with one another. "Ignition, window heat, anti-ice, flight instruments, directors and radios, yaw dampers. Speed brake, flaps twenty-five, twenty-five green stabilizer trim five point nine, start levers idle flight controls," each said to the other, covering everything on the list.[24]

As the Boeing 727 waited on the taxiway, James Watkins started to feel the pressure of keeping all the aircraft under his control flowing safely. Having to keep Atlanta informed of his traffic, as well as advising Welch on his two duties while at the same coordinating en route traffic without any radar, Watkins started to lose his focus. He contacted Addison on the radio and gave him landing instructions. "Three one two one, Sugar," he started. "Cleared over the VOR to Broad River." Watkins stopped, shook his head, and corrected himself. "Correction, make that the Asheville radio beacon, over the VOR to the Asheville radio beacon. Maintain seven thousand, report passing the VOR."[25]

"Three two one, Sierra," Addison responded, acknowledging the transmission.[26]

Watkins knew one mistake made by an air traffic controller could be the difference between life and death. He felt he dodged that bullet by correcting the term Broad River to the Asheville radio beacon. The difference between the radio beacons was huge. The Broad River radio beacon lay to the southwest of the VOR Addison approached. The Asheville radio beacon lay to the northwest of the VOR. With the change in the wind, Watkins wanted to send Addison to the north of the airport over the Asheville radio beacon to land from that direction. If Addison mistook the Broad River radio beacon for the Asheville radio beacon, it would send him right into the path of the departing 727.

How many more bullets, though, could Watkins and other controllers keep dodging before planes began falling out of the sky? The frustration of being an air traffic controller became stronger and stronger with each passing day. He wondered how many passengers would have to die before the FAA realized how insufficient and inefficient the nation's airways really had become. Watkins wiped away his doubts and focused on his job. He knew what to do. He just forgot to explain it clearly to David Addison.

The increased pressure on air traffic controllers was due, in large part, to the austerity policies of FAA chief William McKee.

Air Force General William "Bozo" McKee became the administrator of the Federal Aviation Agency on July 1, 1965. While President Johnson expressed full faith in McKee's abilities, there were senators in Congress who voiced doubts. McKee had a reputation for cutting staff and personnel to make budgets work, which put him at odds with unions. He also had never piloted an aircraft, something almost unheard of among Air Force generals.[27]

Within six months of his nomination, the airways above the United States had become so congested that many within the aviation community expressed fear that the system was broken. At first, with the advent of the first generation of passenger jet aircraft, specifically the Boeing 707 and Douglas DC-8, the airways were not particularly strained by these new fast aircraft. These jets serviced long-haul coast-to-coast routes, with perhaps a stop in St. Louis or Chicago along the way. Only the largest of the airports within the country could handle those aircraft, and congestion was contained.

The second generation of jet aircraft, specifically the Boeing 727, Douglas DC-9, and the British Trident, swiftly expanded the number of jet aircraft in the skies.

Aircraft manufacturers designed these aircraft at the behest of the airlines, with smaller airports in mind. Suddenly, in all regions of the country, it became possible to see and fly in jet aircraft.

However, questions and doubts began to arise within the aviation community regarding the ability of air traffic controllers to handle this large increase in traffic. On February 6, 1966, Clifford Burton, executive director of the Air Traffic Control Association (ATCA), warned that aviation might soon have "an inefficient, inadequately staffed air control system that will either have to handle less traffic or handle present traffic with less safety."[28] His warning could not have been timelier, for in the last six months four 727 aircraft had crashed, resulting in the deaths of 264 people.

During the same period, the Bureau of the Budget issued a halt to grade escalation within air traffic control ranks. This meant frozen pay for the controllers as well as a halt in promotions. Burton also stated that President Johnson was not aware of how these economy directives had brought about the "dangerously low morale that permeates the controller community."²⁹

Johnson, however, had to know. By this point in his term, he had witnessed the unraveling of his Great Society. Frustrated, he watched as the funds for his utopian society were sucked into the cesspool of war. In December 1966, he announced he would ask Congress for an additional nine to ten billion dollars on top of the fifty-seven billion already allotted to the war chest for fiscal year 1967. This would bring the total amount for defense to between sixty-seven and sixty-eight billion dollars.³⁰ There would soon be no money left for Johnson to fund anything but the war in Vietnam.

As a career politician and businessman, Lyndon Johnson knew the relationship between a person's paycheck and morale. It was not unlike the relationship between a campaign contribution and the contributor's access: the larger a donation to a presidential campaign, the more access one had to the winning candidate. A millionaire several times over, Johnson understood money. He also understood who spent the money in government. That was the purview of the United States Congress, and Johnson figured if aviation safety was at risk, Congress should fix it. He had a war to fight. It seemed that the only way Johnson would act on this problem would be if someone he knew, a close friend, advisor, or family member perhaps, got on the wrong plane at the wrong time.

After Burton's warning, McKee announced that he would take some action on it. He did not move quickly enough. On March 9, 1967, a Trans World Airways (TWA) DC-9 collided with a Beechcraft 55 aircraft over Urbana, Ohio, and twenty-six people perished instantly in the nation's crowded sky.

The American public had witnessed midair collisions before. The collision between a United Airlines DC-7 and a TWA Super Constellation above the Grand Canyon in 1956 killed 128 people. In 1960, two other United and TWA aircraft collided over Brooklyn, New York. This time, a speeding United DC-8 jet aircraft struck a slower-moving TWA Super Constellation above the city, sending the DC-8 crashing into a Brooklyn neighborhood, while the Constellation crashed on Staten Island. One hundred and thirty-four people, including six on the ground, lost their lives because of a hurried United Airlines pilot.

It is a stated belief in air accident investigative circles that an "aircraft accident is a startling and shocking occurrence. There is intensive interest in it for a brief time and then it fades from the memories of most individuals concerned until little remains other than the facts recorded in the report."[31] For shock value, the Urbana crash did not contain much voltage. With the war accelerating in Southeast Asia and Johnson keeping his war face painted, public attention turned away from the unsafe skies to the daily uptick of the body count in Vietnam.

Those on the ground, though—those responsible for keeping the flying public safe while in the air—started to raise their collective, concerned voices. Air traffic controllers knew that a problem existed. Politicians raised concerns mainly about the rationality of mixing small, slow general-aviation aircraft with large, fast commercial jets on the same airways. But with the nation's attention riveted on the war and the diversion of funds, Johnson remained relatively silent.

The National Association of Government Employees (NAGE), a union representing eight thousand of the nation's air traffic controllers, had, back in March, predicted the future, and it became more confident that it knew what it was talking about.[32] The word went out from the union that the nation's airways were not safe, and this time the FAA listened, but only because of Urbana.

In the days after Urbana, McKee sent evaluation teams from Washington, D.C., into towers and flight centers around the country. With the accusations thrown about by NAGE, in combination with the Urbana accident, McKee decided that some effort must be made to listen to the complaints. McKee did not like what he heard back from the field. Controllers were scared. They complained that the air control system was strained.[33] This was precisely what NAGE had stated. On top of that, McKee implemented a program forcing psychiatric evaluations upon air traffic controllers. This program allowed the FAA to disqualify those controllers who did not meet their standards without informing them of their rights to "free medical treatment, compensation, or even offering them other jobs."[34] That program, coupled with the fact that the controllers knew they would become pawns in the aviation funding game, sent the teams back to Washington wishing they had never visited the field in the first place.

Acknowledging the low morale of the nation's air traffic controllers, McKee finally ordered that immediate and long-range steps be taken at towers and ATC centers to "maintain and improve air safety."[35] He stated that "these actions are mandatory if we are to keep in step with accelerated flight operations."[36] McKee further ordered a cutback on administrative duties of controllers, a change to make it easier to get more overtime pay as well as an equal sharing of assignments around holiday time.[37]

These efforts did not put any more money into the FAA coffers, so the system continued to operate at inadequate staffing and spending levels.

McKee merely shuffled duties among the FAA personnel, increasing workloads for some and decreasing them for others. He had a tin ear for direct complaints from his personnel: they wanted at least six hundred more air traffic controllers and they wanted them soon, or Urbana would be just the beginning of bumper cars in the sky.[38]

4

"Twenty-two rolling"

At his current speed, David Addison figured he had about two minutes to clear up the confusion of the clearance sent from James Watkins. While he had never flown to the Asheville airport before, he had already pulled out his approach plate and turned it to the ILS approach procedure for the Asheville airport. After all, Atlanta had told him to expect an ILS landing in Asheville, and Watkins had not changed it.

There are four approaches to the Asheville airport. Three of them are from the south and one from the north. On an ILS approach from the east—the course the Atlanta controllers had told Addison to expect—a pilot would fly to the Asheville VOR and then turn southwest towards the Broad River radio beacon; passing that, he would turn onto a heading of 320 degrees towards the airport. On the approach from the north, called an ADF-2 approach, a pilot would fly to the Asheville VOR from the east and then turn northwest towards the Asheville radio beacon. Once there, the pilot would turn south towards the airport.[1]

Per federal regulations, an approach controller must "notify an arriving aircraft at the time of first radio contact, or as soon as possible thereafter, the type approach clearance or type of approach to be expected if two or more approaches are published and the clearance limit does not indicate which will be used."[2] Watkins did not relay that information to Addison.

David Addison had been flying for a long time. He had dealt with controllers before who had a lot less experience than he did, and many of them did not follow prescribed procedures. Only in the rarest occasions would this create even the hint of a problem, but at this moment Addison's problem was that no matter how hard he looked, he could not locate the Asheville radio beacon on his ILS approach plate. He repeated Watkins' clearance to himself: "Cleared over the VOR to Broad River, correction, make that the Asheville radio beacon, over the VOR to the Asheville radio beacon. Maintain seven thousand; report passing the VOR."

Surely Addison asked himself whether the correction was to the name of the beacon in question, or whether it was a completely different beacon at a different location. If a pilot approached Kennedy Airport in New York and the controller cleared him to Idlewild Airport (the former name of that airport), the pilot would not be looking for a different airport. He would understand the controller just used its old or alternate name. If the Broad River radio beacon and the Asheville radio beacon were two different beacons in two different places, marking two different approaches, the controller would have told him which approach to use. The regulations dictated that. Certainly, it had to be the same beacon, because no matter how closely Addison looked at the approach plate, no matter how he read it, he could not find any beacon on the ILS approach chart with the name Asheville.

He could not find it because it was not there. The Asheville radio beacon was not part of an ILS landing, so it was not listed. Still, time remained to clear up the confusion, and Addison knew exactly how he would do it when he approached the VOR.

Waiting in the Piedmont 727, Schulte and Conrad could hear the banter between Paul Snell in his Martin 404 approaching the airport and the tower on the company radio located inside the cockpit. "I'm five miles southwest of Broad River, 1022,"[3] Snell said, advising the tower of his location.

"Piedmont 22, hold short. We got a company Martin on approach just about over Broad River circling for one six. We should have him in sight shortly,"[4] Welch advised the Boeing jet crew.

"Okay, we'll hold it short,"[5] Conrad replied from the cockpit, impatiently waiting to take off. He knew that situations like this caused delays, and since they had already been held up in Atlanta, this became one more delay they would have to explain. "Cancel the damn thing, Paul,"[6] Conrad urged. He wanted Snell to cancel his IFR approach and use contact approach.

"OK. Ten twenty-two would like a contact approach,"[7] Snell stated, having taken his friend's advice. Snell had flown into Asheville many times and knew his location exactly. He knew a contact approach would allow him to deviate from the approach procedures at the airport as long as he kept the airport in sight and there were no interfering clouds or aircraft. The pilot of an aircraft has to ask for this approach. Air traffic control cannot even suggest it to a pilot.

4. "Twenty-two rolling"

"Ten twenty-two. Contact approach," Welch said, approving the approach. "Left or right turn-in is approved."[8]

Suddenly, the three-way conversation between Snell in the 404, Conrad in the 727, and the tower took on comic proportions.

"Now which way will you be turning in, please?" Welch asked.

"We'll come up out of the way of the jet. Is he going to climb straight out on one six?" Snell asked.

"Affirmative. Will you be west of the localizer?"[9] Welch asked, wanting Snell's location in relation to the Broad River radio beacon.

"He's making a right turn on one six, I guess," Conrad chimed in from the jet's cockpit.

"I'll go east," Snell replied.

"Oh, no," Conrad replied laughing. Ignoring the authority of Jesse Welch as air traffic controller, Conrad decided to take over as traffic cop. "Ten twenty, twenty-two it is. We'll turn on off and go direct to the VOR."[10]

The usurpation of his authority did not sit well with Jesse Welch. "Piedmont twenty-two, taxi in position and hold," Welch admonished the interfering crew of the Boeing jet.

"Twenty-two, roger," Conrad replied sheepishly.

"Cancel for ten twenty-two now," Snell stated, taking over approach responsibilities for himself.

"Ten twenty-two roger, canceling," Welch said into the radio.[11] That was one less plane he had to worry about.

"I thought it was going to take you all day, Paul," Conrad chimed in on the radio, laughing and ribbing his fellow Piedmont pilot.

"Well, I'm down here at Broad River, finally," Snell replied, not being able to cancel before he arrived over the radio beacon.

"Attaboy," Conrad replied. "We'll see you, buddy."

"OK," Snell retorted.

"Ha, ha, ha, ha," Conrad added, still laughing and ribbing Snell over his caution. Conrad gripped the throttle in the cockpit tightly, and Captain Schulte sensed his anxiety.

"We're not cleared for takeoff yet," Schulte admonished his copilot.[12]

Jesse Welch now realized he could clear Piedmont 22 safely and get the jet on its way to Roanoke. "Piedmont twenty-two, maintain runway heading until reaching five thousand. Cleared for takeoff," Welch commanded.[13] He watched from the tower as the 727 shuddered for a moment, saw it almost imperceptibly grind down on its rear tires, and then stared in awe as the machine gathered speed and started down the runway like a large bullet seeking early release from a rifle barrel.

"Twenty-two rolling," a relieved Schulte spoke into the radio.

"Ground off," Wilson interjected, confirming the auxiliary power unit had been turned off.[14]

The engines revved louder as the aircraft sped faster down the runway. The crew felt the slight bumps of the runway as the tires rolled even quicker over the concrete. The eyes of the three crewmembers moved rapidly from the instrument panels to the quickly disappearing runway in front of them. The captain had sole responsibility to take the aircraft from the runway into the sky. "Check your rudders," Schulte commanded.[15]

No problems existed with the rudders. No problem existed with anything inside the cockpit on the instrument panels nor outside around the blue sky. Conrad exhaled. "Yeah," he said with the satisfaction of a speed junkie getting an acceleration fix.[16]

A satisfied Schulte pulled back on the yoke, and they lifted into the sky before the rest of the aircraft followed. He pulled back further and felt the plane lift off the ground. "There you go, my friend," he said.[17]

The sheer power of the jet aircraft surprised all who rode in it. The experienced crew knew that the next five minutes of flight were among the busiest and most crucial parts of the trip, right alongside the landing. A constant vigilance of cockpit instruments, as well as particular attention to scanning the sky in front of them for advised and unadvised traffic, would keep the undivided attention of all three crewmembers. While the sterile cockpit rule had not been established yet, it was vital that the crew eliminate any type of distractions from the work at hand, and there was no commercial aircraft in the world that required more of a crew's absolute attention than the Boeing 727.

A span of several months in 1965 and 1966 revealed some ugly truths about the Boeing 727. If pilots became distracted in the cockpit, death could find them and their passengers with startling speed. The Boeing 727 entered commercial service on February 1, 1964. Within the first eighteen months of its existence, the tri-motor jet performed flawlessly, obtaining a perfect safety record. Then, inexplicably, in August 1965, one crashed while on approach to Chicago's O'Hare Airport. In November 1965, two more crashed while on approach to other airports, and just over two months later, another one crashed into Tokyo Bay while attempting to land. These four crashes took a total of 264 lives within that six-month period; many in the aviation community, in the traveling public, and, even worse for the industry, in the United States Congress demanded that the 727 be grounded.

The first deaths occurred on August 16, 1965, less than two months after "Bozo" McKee took over the FAA. Just before 9:30 p.m. a United Airlines Boeing 727 dove into Lake Michigan while on approach to O'Hare. All thirty individuals aboard the aircraft perished in a crash so violent that investigators never found the cockpit voice recorder (CVR). While investigators never determined the exact cause of the crash, some believed that the pilots misread their altimeter system by ten thousand feet.[18]

After the typical horrified reaction by the public, the accident receded into history, and after a few days the traveling public forgot the crash, forgave the aircraft, and continued flying on it. Then, on November 8, 1965, at 7:05 p.m. another 727, this time an American Airlines jet, crashed near the banks of the Ohio River while attempting to land at the Greater Cincinnati Airport. Out of the sixty-two aboard the aircraft, only four survived. Civil Aeronautics Board investigators determined that the pilots of the aircraft caused the accident by failing to monitor the altimeters during their approach into Cincinnati.[19]

On November 11, 1965, just before 6:00 p.m. before the traveling public could even digest the horror of the Cincinnati crash, another United Airlines tri-jet, the third 727, crashed just short of the runway in Salt Lake City, Utah. Out of the ninety-one passengers and crew aboard, over half survived, but forty-three died in the crash. Ironically, the post-crash fire, not the impact, killed all forty-three, as the landing gear popped up into the fuselage and ruptured the fuel lines located on the bottom of the aircraft. Safety investigators blamed the accident on the captain of the aircraft, who allowed the plane to descend at three times the safe descent rate.[20]

Within hours, calls to investigate the 727 resonated throughout the United States. In the United States Congress, Senator Vance Hartke of Indiana demanded that investigators probe the aircraft and issue a report on its airworthiness. Even President Johnson expressed concern regarding the aircraft. He called on the FAA and the CAB to send their best men to investigate.[21]

Within hours of the accident, CAB and FAA officials expressed complete confidence in the aircraft. They also acknowledged that no pattern existed amongst the crashes and that they had no idea what had caused the planes to crash. Through their spokesman, CAB officials diverted blame away from the aircraft. The statement read, "The jet model had achieved nearly 400,000 hours of successful operations in scheduled air service before the accidents. It passed very rigid certification tests of the FAA before it was put into service and nothing has turned up in our investigation to cause us to doubt its stability."[22] Plainly, this was not the truth. Anyone who could tell time quickly realized that all three accidents occurred at night and while they were on approach to their respective airports. Privately, aviation experts worried that 727 pilots were experiencing problems controlling the descent of the aircraft.

With congressmen screaming for the aircraft's grounding, and the nation's newspapers covered with pictures of burnt-out 727s, the FAA became quietly concerned about the stability of the aircraft. The reputation of the FAA, as well as Boeing and the CAB, depended upon the exoneration of the jet. If, after tens of thousands of hours' testing and certifying an aircraft, investigators revealed the 727 was not safe to fly, the FAA would have a public-relations nightmare on its hands. Senator Hartke demanded that the CAB issue a public report on the safety of the aircraft.[23] The CAB and the FAA ignored him; that is, until February 4, 1966, when an All Nippon Airways 727 crashed into Tokyo Bay at 7 p.m. while on approach to Haneda International airport. One hundred and thirty-three people lost their lives, and thousands of miles away in Washington, D.C., the CAB and the FAA decided they could not ignore the concerned senator any longer.

Administrator McKee knew he had to act quickly. The Japanese crash almost doubled the number of people killed in 727s during the past four months. He knew that United Airlines, owners of two of the lost aircraft, had begun retraining its pilots on the 727 with an emphasis placed on the plane's sink rate. While releasing a statement through a spokesman that he "saw no reason to ground the aircraft at this time,"[24] McKee urgently summoned representatives of all the airlines in the world using the 727 to Washington to discuss the accidents. Knowing he had no time to waste, he set the meeting for the next day.

Seventeen airlines, all those using the 727, sent representatives to the unprecedented meeting. After an all-day meeting, the exhausted representatives of the airlines and the FAA agreed to propose changes in crew procedures while flying the 727. The proposed changes addressed the problem of the rate of descent during landings. After months of denial, Clifford Walker, deputy director of flight standards for the FAA, admitted that there was a common thread weaving through the accidents: each occurred at night and on approach with relatively good weather. He added that the airline representatives called to the meeting regarded the 727 as a "wonderful" plane and, protecting his agency as well as the department he led, that intensive investigation revealed no flaws in the design of the plane.[25]

Changes in cockpit procedures included reporting altitude levels more frequently as the plane approached an airport. Callouts of speed and sink rates would be added the closer the aircraft came to the ground. The FAA also proposed new training maneuvers to emphasize sink rates, thrust requirements, and flight instruments.[26]

Decisions made by the FAA administrator affected the flow of hundreds of millions of dollars. The abrupt grounding of the 727 would certainly not promote commercial aviation, a mandate of the FAA. The airlines owning 727s would lose thousands of dollars a day if the FAA ordered the 727 grounded. Boeing would be out millions of dollars, as 471 of the three-engine jet had already been ordered by the airlines and 224 had already been delivered. The awesome reputation of all Boeing aircraft would be affected by a ruling detrimental to the 727. The void would gladly be filled by foreign aircraft manufacturers already subsidized to the maximum by their own governments. They would be able to fill the void quickly and without any cost to their companies, as the taxpayer paid for their aircraft anyway. A harmful ruling on the aircraft would also damage the credibility of the FAA, the one organization responsible for making sure new aircraft met the standards of commercial aviation. The FAA was not about to indict itself; it was much easier to blame a dead pilot.

During the public hearing on the Cincinnati crash a few months later, the CAB incorporated a public report on the safety of the 727 into the final report of the crash itself. The CAB investigators concluded that the "specific investigation of the Boeing 727 flight performance and characteristics revealed no design deficiencies or unsatisfactory operating characteristics."[27] They did admit, however, that the 727 "*does have* highly responsive and versatile flight characteristics and that these favorable characteristics may be misleading to the pilot, or are presenting the impression that greater liberties may be taken with the aircraft in normal operating situations...."[28]

As to crews flying the 727, the CAB gave this warning: "The Board must re-emphasize that the responsibility and authority committed to an airline captain requires the exercise of sound judgment and strict adherence to prescribed practices and procedures. Any deviation can only result in a compromise of aviation safety. Airline management, too, has a heavy responsibility for devising, developing, and implementing methods of procedures designed to insure that all of their pilot personnel constantly exercise a conservative, prudent, approach to their daily work."[29]

Senator Hartke finally got his public report. The FAA, the CAB, Boeing, and the airlines could not endure another fatal 727 crash blamed on the flight crew. If blame landed on the aircraft itself, the pressure from the public and Congress would be so great that the plane would probably have to be grounded. The Lockheed Electra scenario had returned to haunt the CAB and the FAA.

By the late 1950s, aircraft manufacturers at Lockheed had developed a truly transitional airplane. The engineers called it a propjet. Its official name was the Lockheed Electra L-188. It was an aircraft driven by propellers but with the speed and capability of a jet. Many airlines enthusiastically placed orders for this new craft to help them service regional airports that did not have jet capability and wanted faster speeds than a purely propeller-driven aircraft could offer. Lockheed had truly combined a jet and prop plane into one aircraft. Pilots raved about its features, and the aircraft became popular with the traveling public—that is, until the wings started falling off in flight.

On September 29, 1959, just thirteen months after the certification of the L-188, a Braniff Electra, only eleven days old, lost its left wing over Buffalo, Texas. As the left wing tore off, the plane fell like a dove hit with birdshot and disintegrated upon impact with earth. Debris from the aircraft rained across the farmland of eastern Texas, the left wing landing a mile away from the main wreckage. Strewn between were the remains of thirty-four passengers and crew. For CAB investigators, the wing became the first clue to solving the accident.[30]

After months of investigation, the air sleuths from the CAB were no closer to solving the tragedy than they were on the first day of the crash. From the location of the wing, they knew it had fallen off first, but the question remained: why?

As the investigators mulled over the frustratingly small number of clues left, another Electra crashed over Tell City, Indiana, on St. Patrick's Day, 1960. Its left wing, too, had torn off in flight. Aboard Northwest Flight 710, all sixty-three people perished when the plane plunged into the earth at over 600 miles per hour. The impact created a hole in the Indiana ground sixty feet deep. The high-velocity impact with earth destroyed any chance of recovering much from the fuselage. Almost two miles away, though, investigators found the right wing of the aircraft intact enough to perhaps reveal the cause of the crash.[31]

On top of quickly solving these mysteries, the CAB and the FAA had to decide whether or not to ground the Electra L-188. Two brand-new aircraft had their wings torn from their fuselages, resulting in the deaths of almost one hundred people, and no one knew why. At this point, there were 113 Electras carrying over twenty-five thousand people per day around the United States. Officials had to weigh the risk of the passengers on the aircraft against the economic impact grounding the Electra fleet would have on air commerce. The CAB had the responsibility of recommending the grounding of aircraft if they were deemed dangerous, but only the FAA could order the grounding of an entire fleet.[32]

4. "Twenty-two rolling"

General Elwood Richard "Pete" Quesada received his appointment from President Dwight Eisenhower to lead the FAA in 1959. His successful use of tactical air power in support of ground troops during World War II made him a war hero and contributed greatly to the defeat of Nazi Germany. After the war, he became a vice president of a major aircraft manufacturer, and then Eisenhower picked him to be his special assistant on aviation matters until his appointment to the FAA.

The decision to ground the Electra fleet rested solely with Pete Quesada. "If I accomplish nothing else," he once said, "I hope to make the entire FAA staff conscious of our primary obligation to serve the public interest."[33] His ability to serve the public interest in the Electra matter was up for debate. The major aircraft manufacturer of which he was vice president just a few years earlier was Lockheed, the makers of the Electra.

Quesada knew there was a problem with the wing of the aircraft. He quickly made a decision, to be effected immediately, that included restricting the airspeed of the L-188 to 225 knots. He ordered every wing of every Electra to be inspected. He forbade the crew to use the "automatic pilot" mechanism on the aircraft, and he ordered Lockheed to figure out the problem with its airplane, but he decided to keep the aircraft flying.[34] He stated,

> Grounding it would have been the easy way. I don't think anyone could have possibly criticized me if I had done it. But I was not only certain the wing would be safe at slower speeds, I was certain it would be the strongest wing in the whole American fleet. By reducing speeds to 225 knots, we were automatically increasing its ability to bear stress by 50 percent. I don't think any man in my position should act out of self-preservation and self-preservation was the only reason I had for grounding. I don't think the director of any government regulatory body should be allowed to harass the industry he regulates. I felt grounding, under the circumstances, would have been harassment.[35]

Pete Quesada sided with his former business partners at Lockheed over the American flying public.

The five members of the CAB had a different outlook on the Electra from Pete Quesada's. During the Quesada-mandated inspections of the Electra wings, airlines found sheared wing clip rivets in over 90 percent of the aircraft. John Pahl, chief of the CAB's safety bureau's engineering department, recommended that the board "recommend the grounding of all the uninspected Electras immediately."[36] Secretly and unanimously, the board voted to follow Pahl's recommendation. The board members, again secretly and confidentially, met with Quesada and gave him their opinion. Politely, Quesada turned them down, and the Electra kept flying.[37]

Nothing remains a secret in Washington for very long. Within twenty-four hours of the secret meeting, Senator Hartke phoned Quesada. Having lost close friends in the Indiana crash, he had a personal interest in the accident. He asked Quesada if it was true that the CAB had recommended the grounding of the Electra fleet. Quesada replied that it was true. When pressed by the senator about what he was going to do regarding their recommendation, Quesada replied that he would do nothing. The planes would keep flying. Hartke held a press conference and revealed, "I am shocked that General Quesada would take the chance of ignoring this advice and risk further death from these airplanes."[38]

In response to Hartke's statement, Quesada ordered that additional measures be taken to find the problem with the Electra. In his article "Brilliant Detection in Jet Age Mystery," Paul O'Neill wrote the following about Quesada's decision:

> He ordered Lockheed to completely re-audit all the mathematical calculations involved in its original planning of the Electra. He ordered that Lockheed conduct a tremendously sophisticated flight test program under the most extreme conditions of turbulence and to engage in static testing and wind tunnel experiments far more complex than those originally done on the plane.[39]

Quesada also sent a team from NASA to the Lockheed plant to help their engineers, and Douglas and Boeing also sent personnel to help Lockheed solve the mystery of the broken wings.

The combined efforts of all these groups paid off when half a year later, engineers discovered what caused the wings to fall off the Electra. Paul O'Neill summed up the situation in his article:

> As long as the Electra's outboard nacelles retained the stiffness for which they had been designed, the plane was safe and sound. But if the struts and braces inside the nacelle structure cracked or otherwise loosened, and if the wing was jolted hard by air turbulence or by a sudden pull up, and if the plane was moving at a high speed at the time, a curious chain reaction effect could be induced.... The propellers turned at 1280 revolutions and the turbine at 13,280 revolutions a minute. Thus when running they comprise a huge gyroscope.... When the sudden jolt caused the wing to flex and agitated a nacelle, this whole package of spinning metal, which projects far out in front of the leading edge of the wing, would begin wobbling in its mount, like a pointed finger with its tip describing a small circle.... The oscillating wing and the swirling power plant suddenly begin exchanging energy.[40]

The mystery had been solved. The exchange of energy would cause the wing to snap off, and the Electra suffered the same fate as most other aircraft when mechanical flaws are spotted so soon into its existence. The public lost confidence in the aircraft, and no more were made.

With Quesada's decision to keep the plane flying proven correct, the CAB faced some surprising criticism. O'Neill wrote at the time, "It seemed to suggest that the board was incapable of any genuine concern for the public at all and that an airplane accident was lamentable, not so much because it killed people as because the aircraft at fault demeaned itself and caused fearful embarrassment to the industry."[41] CAB investigators etched the fallout from the Electra issue in their memory and kept it with them through their incorporation into the NTSB. Never again would they recommend the grounding of such a large fleet of aircraft.

If the management of Piedmont Airlines knew about the controversies surrounding the handling of the Boeing 727 aircraft in general, no one paid particular attention to them. Their specific aircraft, N68650, *Manhattan Pacemaker*, had enough problems of its own without even worrying about its in-flight handling capabilities. With all its individual problems, it is amazing that Piedmont management let the aircraft fly at all.

Throughout the 1960s, midair collisions had been one of the most pressing problems confronting air safety officials. Roadway collisions for aircraft, though, were not. On May 25, 1967, as N68650 was being pulled to a hangar to be worked on, a car struck the 727 on its belly. As a flagman watched in horror, frantically waving his arms, the driver of a car failed to see that the huge aircraft blocked the roadway and drove underneath it, causing $25,000 in damage to the jet. The top of the car was flattened, and the *Manhattan Pacemaker* retired to the hangar for a few days.[42]

After management placed the aircraft back in service, two other incidents took place that further weakened confidence in the aircraft. As the jet approached Smith Reynolds Airport in Winston-Salem, a warning light flashed in the cockpit, indicating that the nose wheel had not locked into its proper place. As the crew took the aircraft down closer to the airport, witnesses on the ground could see that the wheel was in its proper position. After the crew had briefed the passengers aboard the aircraft on how to brace themselves in a crash position, the plane landed safely, surrounded by rescue equipment.

A quick check of the aircraft by Piedmont mechanics did not reveal any malfunctions with the aircraft, so they released it on to Asheville. As the plane prepared to land there, the warning light again flashed and the plane turned back to Winston-Salem for a further maintenance check. Flanked by rescue vehicles, it landed safely again as mechanics struggled to find the problem with the landing gear lights.[43]

A larger and more lethal problem plagued the *Manhattan Pacemaker* than the unreliability of its landing gear lights. The RCA weather radar installed on the aircraft did not work properly; indeed, it had not worked properly for most of the months of June and July. This unit possessed the ability to detect foul weather in flight, allowing the cockpit crew to divert around the problem area and continue safely in flight. Without operable weather radar on its aircraft, the affected airline put the lives of the passengers and crew at risk.

Turbulence had always lurked in the sky as a silent killer. With jet aircraft, the phenomenon became more lethal because of the shortened pilot response time with high speed jets. On February 12, 1963, a Northwest Orient Boeing 720 (a shorter version of the 707) took off from Miami International Airport. A few minutes after departure, the aircraft entered an area of turbulent weather that shook the aircraft violently until the wings sheared off and it shattered in the swamps of the Everglades, killing all forty-three people aboard.[44] On December 8, 1963, a Pan Am 707 in a holding pattern over Elkton, Maryland, exploded in flight as a bolt of lightning struck the aircraft, sending all eighty-one aboard to their deaths.[45] In the early morning of February 25, 1964, turbulence struck yet again as it shook an Eastern Airlines DC-8 until it plummeted into Lake Pontchartrain in New Orleans, Louisiana. All fifty-eight aboard died instantly.[46]

Even though there were over twenty other turbulence-related accidents since 1960, the government hesitated to admit that turbulence even posed a problem to aircraft. Throughout the invisible reign of turbulence's terror in the early '60s, the FAA and the CAB only warned aircraft that the possibility existed for bad weather and recommended crews slow their speed to 325 miles per hour while navigating turbulent skies. The director of air safety for the CAB even went so far as to deny a turbulence problem. In an interview with LIFE magazine, Bobbie Allen stated, "If I were trying to write a best-selling novel, I'd put turbulence in the title. There seems to be so much interest in it. We don't know if we have a turbulence problem or not."[47]

Thankfully there were others at the highest levels of government who did see a problem with turbulence. By the end of December 1964, the FAA mandated, "No person may operate any transport category airplane or a non-transport category airplane certificated after December 31, 1964, unless approved airborne weather radar equipment has been installed in the airplane."[48] The FAA also stated that the radar must be "in satisfactory operating condition," and if it was not, the aircraft could not be dispatched in instrument flight rules (IFR) or nighttime visual flight rules (VFR) con-

4. "Twenty-two rolling" 43

ditions if "thunderstorms, or other potentially hazardous weather conditions that can be detected with airborne weather radar, may be reasonably expected along the route to be flown."[49]

By 1967, attitudes had changed regarding airborne weather radar equipment. Many pilots knew that that inoperable weather radar could seriously jeopardize the safety of commercial air passengers and the crew themselves. Several Piedmont pilots knew it, but it seems that the management at Piedmont Airlines did not see the problem the same way.

On June 20, 1967, a pilot at Piedmont revealed the big, dark secret about the *Manhattan Pacemaker* in its maintenance log: the aircraft's radar did not work at all. In fact, the pilot wrote, "Radar unfit for weather detection."[50] On June 23 another pilot wrote, "Radar absolutely useless."[51] Maintenance did respond, checking the wiring, and certificated that the radar operated correctly. Two days later, another pilot bluntly stated what others had been thinking. He wrote, "This airplane should be grounded until an operable radar is installed. Present set is unfit for weather mapping."[52] Management, however, did not ground the aircraft; they kept it flying. Maintenance replaced several parts and accessories in the aircraft, but the problems continued. On June 29, pilots continued their indictments. "Radar is by no means up to Piedmont standard. To obtain even a legal picture of a hairy t-storm requires very tedious tuning and if distance range is changed, requires a new tune job. If radar is of any use at all it requires one man to tune almost constantly."[53] Maintenance responded that the unit checked normal but did recommend different tuning procedures. A few days later, a pilot wrote in the log, "No change in radar with new tuning procedure."[54] Yet management kept piling passengers on this airplane, placing them in an aircraft that was virtually blind beyond the human vision of the aircraft's pilots.

Crews, though, could be briefed on the weather along their route. Pilots had contact with Weather Bureau stations around the country, but there were not many left. FAA flight service stations around the country also offered weather information, but the FAA curtailed these facilities in another round of cost-cutting measures. It also slashed the number of weather broadcasts in other areas.[55] With the FAA cutting costs at every level of aviation, the importance of airborne weather radar increased daily.

As the second week in July rolled around, pilots of the *Manhattan Pacemaker* still flew blindly. On July 13, a pilot summarized his frustration in the maintenance log. "Radar remains the same. The radar complies with the reg only and is by no means up to Piedmont standards; a dependable radar is not only desirable but is fast becoming a necessity. No change in radar, still useless."⁵⁶ By that time, Piedmont's standards had blurred. They kept the blind plane in the sky.

Whether someone at Piedmont notified the FAA regarding the absence of operable radar on their Boeing 727 is not known. The FAA did, though, have an office at Smith Reynolds Airport in Winston-Salem. Guy T. Crowe headed the office as the FAA inspection supervisor of the Winston-Salem Air Carrier District Office. The personnel within that office possessed FAA surveillance responsibility over Piedmont airlines. Mr. Crowe summed up his responsibilities towards Piedmont in this way:

> As Supervisory, we have certain district boundaries in which we have general air carrier safety responsibilities and functions. Our primary function is the administration of the Piedmont Air Lines operations certificate and we have a program of surveillance in which, in effect, is a sampling of the Piedmont operations and maintenance, to check their continued compliance with the regulations, standards.⁵⁷

As Piedmont airlines brought the 727 into its fleet, management had to be aware of the design flaws in the aircraft. These flaws had killed over two hundred people within a four-month period just a year before, almost causing the aircraft to be grounded. With the individual problems facing the *Manhattan Pacemaker*, as well as the inherent problems with the design of all 727 aircraft, Guy Crowe's main focus should have been educating Piedmont management and pilots regarding the special characteristics of this aircraft. He should have had his inspectors constantly observing the operation of this aircraft so that Piedmont would not suffer the crashes that other airlines endured when they first flew the 727. He should have acted upon the concerns and write-ups regarding N68650's weather radar unit. However, neither Guy Crowe nor any of the Winston-Salem inspectors were even certified to fly any Boeing 727 jet aircraft. In fact, Crowe did not possess a license to fly a jet aircraft at all.⁵⁸

5

"Somebody got an ashtray on fire?"

With Piedmont Flight 22 off to Roanoke, Piedmont 1022 approaching the airport from the south, and 3121S approaching from the west, along with various other aircraft coming and going, Watkins felt the pressure of his job increase.

With no radar to help him guide the planes, he hoped everyone had heard his orders correctly and clearly. So many lives depended on concise and clear communication both to and from the control tower that one misunderstanding could result in the deaths of many people. Even with his confused clearance to Addison, mixing up the Broad River radio beacon and the Asheville radio beacon, he assumed all aircraft were moving about the sky safely. As soon as 3121S reported over the VOR, Watkins would have Brown lift the takeoff restriction placed on Flight 22 and send it to the left towards Virginia. Watkins picked up the phone with the direct line to Atlanta air traffic control to advise them of the position of the aircraft under his control.

"Atlanta," the other end answered.

"Asheville Piedmont ten twenty-two is out of seven…," he started, but Addison interrupted him with a position report from his Cessna.[1]

Addison had spent the last two minutes of flight time searching for the Asheville radio beacon on his ILS approach plate. He may even have handed the plate to Bob Anderson to see if he could find it; if so, Anderson could not find it either. With all his flight experience, Addison had a fix for the problem, and he would tell the controller once he passed over the VOR. Addison's instruments indicated they were now directly above the VOR. He radioed Watkins, the transmission somewhat garbled, "Two one Sierra, just passed over the VOR. We're headed for Oh Good Shit, via 238 for, uh, Asheville now."[2] Since Watkins' earlier approach clearance did not state which direction to head in, Addison figured he would clear up the confusion and tell Watkins precisely what direction he would approach from the VOR. It was a sound plan, but unfortunately, Watkins never heard a word of it.

Watkins continued talking to Atlanta as Addison relayed his message. "And, uh, put two one Sugar three one two one Sugar out of seven and Piedmont twenty two be off at five nine," Watkins told the Atlanta controller as Addison finished.[3]

The Atlanta controller wrote down the information and filed it.

"Two one Sugar, roger by the VOR. Descend and maintain six thousand,"[4] Watkins replied to Addison's transmission. Watkins had waited for Addison's report for two minutes. He thought he knew what Addison would say, so there had been no reason for him to listen; after all, he had told Addison to report over the VOR. Watkins felt it more important to complete the conversation with Atlanta. James Watkins failed to realize that he had just authorized David Addison to fly right into the flight path of Piedmont Flight 22.

"We're leaving seven now,"[5] Addison replied, confident that he had cleared up the matter of the approach with Watkins. The controller had not told him otherwise, so he continued 238 degrees to the southwest off the VOR towards the Broad River radio beacon, not the Asheville radio beacon.

As the Boeing 727 climbed higher into the sky, the captain, first officer, and flight engineer busily engaged themselves with their cockpit duties. They all knew that takeoff, like landing, was the busiest part of a flight, and it required their undivided attention. With Conrad flying the plane, Schulte concerned himself with his duties, including monitoring the cockpit instruments and maintaining a view outside the cockpit window for other aircraft. Even though they were flying under IFR, the pilot not flying the aircraft still had a responsibility to maintain a lookout outside the aircraft, especially in VFR weather. The flight engineer engaged himself in monitoring his instruments and, when he had a free moment, glancing outside the cockpit window.

"Gear up,"[6] Schulte said as the wheels locked safely beneath the aircraft.

Paul Snell's voice boomed over the radio, interrupting the crew's concentration. "Tom, I'm coming up the highway. We're down low. We'll be out of your way,"[7] Snell told him, as the crew of each aircraft continued to work out their own version of aircraft separation.

5. "Somebody got an ashtray on fire?"

"Righto,"[8] Conrad replied to Snell, as if he needed a further distraction from his duties.

"Thank you, buddy,"[9] Schulte chimed in.

Captain Schulte scanned the cockpit instruments again, checking the flaps, the altimeter, and the airspeed. Then, disregarding anything he may have heard about the Boeing 727's handling capabilities, ignoring the written admonition of safety experts at the CAB about the need to pay full attention while flying this type of aircraft, at the most critical time of flight, when total concentration was critical to safety, Ray Schulte decided he needed a cigarette.

"Which highway?" Conrad asked. "I guess that thing just over there," he answered himself, looking at Interstate 26 approximately four thousand feet below them.

"Yeah," Schulte replied, lighting his cigarette. "Climb on out to five thousand, eastbound."

"Five?" (Spoken by a member of the crew not identified in the transcript.)

"Climb power,"[10] Conrad said, pouring on the power to the engines and forcing the plane to climb faster into the sky.

Evan Brown had finally returned from his bathroom break in the Asheville tower. Watkins relayed to him his information regarding Addison's Cessna 310; comfortable the two aircraft were nowhere near close to each other, he told Brown to release Flight 22 from its traffic restriction.

"Piedmont twenty-two, climb unrestricted to the VOR. Report passing the VOR,"[11] Brown said over the radio.

As with David Addison a few moments earlier to the northeast, the controllers in the Asheville tower placed Ray Schulte in a quandary. His altimeter read four thousand, two hundred feet. Even though the controllers had just released him from their traffic restriction, the crew was still eight hundred feet below the FAA published departure restriction placed on them—and every aircraft flying out of Asheville to the south. Every pilot knew that FAA published procedures had to be followed. Federal Aviation regulation Section 91.87 read: "No person may operate an aircraft taking off from an airport with an operating control tower except in compliance with the following: Each pilot shall comply with *any* departure procedures established for that airport by the FAA."[12]

Tower personnel had no firm idea of Schulte's altitude; only he and the other crewmembers knew that. If he continued to climb to five thousand feet, which would take just a few seconds, he could then make his turn and be in compliance with both his traffic and departure procedures. However, he decided to ignore the FAA restriction and make the left turn towards the VOR.

"OK. Unrestricted to the VOR,"[13] Schulte responded. Now they were finally free of any restrictions and, literally, had the whole sky to themselves.

"I still don't see Paul," Schulte said, looking out the window and scanning the sky for Snell's Martin 404.

"He'll be right on down there about five miles on down the highway there, Ray," Conrad said, nodding his head in the direction of Interstate 26 below them. "He usually comes about what we call coming up the highway."[14]

Four miles behind Flight 22 in the Asheville airport tower, James Watkins watched the Boeing727 in a shallow left turn. "Asheville," Watkins answered an unidentified voice over the radio.

"What's, uh, have they got some kind of convention up there today?" the voice asked.

"No, not that I know of," Watkins replied.

"That's the busiest I've seen that airport in a long time," the voice said.

"It's, uh, one extra Piedmont flight. I think they've got some kind of camp, some camp children," Watkins explained.

"OK, thank you."[15]

Watkins remembered Addison's Cessna, which he figured was flying towards the Asheville radio beacon. "Cessna three one two Sugar, cleared for an ADF two approach to runway one six. Report the Asheville radio beacon inbound."[16]

David Addison shook his head in disbelief. For the last three and a half minutes he had expected an ILS approach to the Asheville airport. Even after the confusing clearance when Watkins mixed up the Broad River radio beacon and the Asheville radio beacon, Addison thought he had rectified the problem by stating what dial off the VOR he had turned. Now he had just been given a completely different clearance. He opened his approach plate and took out the ADF-2 plate and quickly scanned it. He complied with Watkins' changed clearance and now turned the plane north towards the Asheville radio beacon.

Back in the cockpit of the *Manhattan Pacemaker*, Schulte and Conrad were still discussing the location of Paul Snell. "Well, I'll climb out of there. He'll be down low," Conrad stated, continuing his climb.

"I'm going to get rid of your flaps. They're not doing you any good at all," Schulte told Conrad, smoothing out the wings to increase a smoother air flow.

5. *"Somebody got an ashtray on fire?"* 49

"Well, I was getting my speed up to it there," Conrad replied.

"Would kind of like to see that thing, bring 'em up on the green,"[17] Schulte told Conrad. Once the green appeared on the instrument panel, the flaps would be where Schulte wanted them.

Larry Wilson listened to the cockpit banter between Schulte and Conrad. He had monitored his instruments throughout the takeoff operation and occasionally glanced out the cockpit windshield to help scan the sky for any planes. Something didn't smell right, and Wilson quickly jerked his head around. "Somebody got an ashtray on fire?"[18] he asked.

Both Schulte and Conrad took their focus off their instruments and from scanning the sky. Smoke was coming from Schulte's ashtray. "I do, I think,"[19] he said.

"You know it couldn't be me," Conrad chimed in.

"Ashtray isn't on fire. That's just the cigarette that's on fire," Schulte corrected. He closed the ashtray lid to extinguish the smoldering butt.

"I'm sorry. I fucked up again, didn't I?" Wilson replied.

"Just for that I burn your damn steak,"[20] Wilson responded, talking about their planned meal for this evening.

Although the whole episode had lasted only twenty seconds, during those twenty seconds, had any of them been scanning the sky, they might have seen a red Cessna approaching them, getting bigger and bigger by the second.

"Twenty-one thousand, we got unrestricted?" Schulte asked, trying to regain control of his cockpit.

"Yes, sir," Conrad answered, sensing Schulte's desire to restore discipline.

"I guess he wants about one six, doesn't he?" Schulte asked, inquiring about the altitude to hit when they get to the Valdese intersection.

"I expect he does. He didn't say it thataway."

"Piedmont twenty two is...."[21] At one minute and eighteen seconds past noon, David Addison looked up from his approach plate and gaped in horror as a huge white and blue Piedmont Boeing 727 bore down on him and his aircraft. He pulled up hard and to the right hoping to avert what he knew would be a mid-air catastrophe. He was too late to avert a collision. His last view was that of his left wing piercing the underbelly of the 727 like a knife filleting a large fish. The Cessna exploded and disintegrated against the outside of the passenger liner, and Addison, along with his two traveling companions, lost consciousness as they fell to the earth six thousand feet below.

6

"A LARGE BALL OF SMOKE AND FLAME"

Western North Carolina (WNC) had remained immune to the surge of civilization right up until the English government bestowed land grants on those brave enough to move into the territory then inhabited by the Cherokee Indians. As the white Europeans inched closer to the Cherokee land in the late eighteenth century, the danger increased. Finally, in a burst of violence, the Indians massacred the first of the interlopers and whooped and hollered as his family fled.[1] The man's family returned to claim the body and seek vengeance on those responsible; the cycle of violence repeated itself for forty years until the Cherokee were placed on the "Trail of Tears" and shuffled off towards Oklahoma.[2]

After the American Revolution, an era of peace fell over the region, in towns like Asheville and Hendersonville. Even as the Civil War raged all around, WNC remained unscathed. As the nineteenth century drew to a close, George Vanderbilt, one of the richest men in the world, identified WNC as a place he wanted to live, so he embarked on building a private palace known to local residents as the Biltmore House. Hundreds of workers, artists, and laborers found their way to Asheville, seeking employment by Vanderbilt to help build his palace. After its completion, many of the workers loved the area so much they decided to stay.[3]

The area increased in popularity as many along the East Coast of the United States fled to the mountains of WNC to escape the oppressive summer heat of the lowlands along the coast. Like those before them, many stayed, and small towns and villages popped up in the area all the way up to the border with Virginia to the north and Tennessee to the west. Lacking the transportation hub to grow large industries, WNC traded on its beauty. Summer camps popped up on the hillsides of the

6. "A large ball of smoke and flame"

Blue Ridge Mountains, beckoning those from the cities to come taste nature.[4]

With the camps came counselors with their eclectic tastes, and they introduced the area to popular culture it had not experienced. The strains of Jimi Hendrix and the Doors slowly drowned out the twangs of Hank Williams and Patsy Cline. By the middle of the 1960s, outside influences started dragging WNC, and its population centers, like Asheville and Hendersonville, into civilization and, in turn, exposed the area to problems it did not want to face.

Racial unrest had enveloped many parts of the country by the summer of 1967. While Hendersonville remained untouched by outright racial protests, residents of the town, both black and white, knew they would soon have to reevaluate their views of each other. The Vietnam War split the country in half. Protests grew larger and louder every day as citizens marched to stop the killing in Southeast Asia. Citizens in Hendersonville and in other towns around the United States began doubting President Johnson's policies in Vietnam. The killing of American troops increased every day, and there appeared to be no end in sight to the butchery. While Hendersonville residents may not have liked having to change their deeply ingrained attitudes, they knew that the wave of change could not be stopped and reluctantly embraced the new thinking, the new culture, the new technology sweeping the nation.

The new Asheville airport (AVL) became the staging area for the arrival of the outsiders. Replacing the aging and almost obsolete A&H Airport, AVL opened in 1961 three miles to the west of A&H. Flight 53, a new F-27 flown by Piedmont Airlines, inaugurated service at AVL.[5] Built away from the mountains to ease navigation, AVL had one runway as compared to A&H's two. The runway, however, was much longer than the former ones and able to accommodate the larger aircraft being bought by the airlines. Runway 16/32 at the new Asheville airport measured 4,100 feet, long enough to suit even the new jet aircraft now in use among the airlines.[6] Equipped with a control tower, the airport had no radar installed because it did not meet the required number of takeoffs and landings. The ILS system, though, equipped aircraft to take off and land even in bad weather, and aviation officials considered an ILS system adequate to safely serve an airport like AVL.

The airport also increased WNC's ability to process a large number of incoming and outgoing passengers. With the influx of campers during the summer, the airport's location next to Interstate 26 allowed easy access to and from the camps. Paul Snell's Martin 404, Flight 1022, alone held thirty-seven children from Florida heading to the Blue Star Camp in Hendersonville. It took the waiting staff members less than an hour from the time a plane landed to get their charges to their respective camps.

Camp Pinewood became a favorite of the many children who came to WNC for the summer. Located just a few miles from downtown Hendersonville and adjacent to Interstate 26, the camp was insulated against the town and traffic by lush green trees. Within the camp, a large freshwater lake beckoned, enabling the campers to enjoy all kinds of water sports, from skiing to canoeing and fishing. An archery range sat on the edge of the camp near large trash heaps that were Camp Pinewood's dump. Game nets hung from trees like large spider webs, enabling the campers to enjoy volleyball and badminton.

Pinewood epitomized the sort of camping adventure youth from around the country could experience in the mountains of North Carolina. It shielded the campers and counselors from the outside world and for a few weeks allowed them an existence uninterrupted by war and racial unrest. It freed them from the confines of their television screens and record players, allowing them to enjoy the pleasures of the outside away from the mind-numbing technology that tempted them at home. Camp Pinewood was a slice of heaven detached from the outside world. It held back the unremitting march of modernism from youth reveling in their innocence, but it could not protect them from two crippled planes tumbling toward the ground.

Harry LaViers, president of the South-East Coal Company in Paintsville, Kentucky, was staying at the Holiday Inn across I-26 from Camp Pinewood. He and his wife were visiting their granddaughter, who was staying at a nearby camp. As he entered his room to take a phone call, he heard his wife cry out from poolside that an airplane had just crashed across the highway. As his wife finished talking, he heard a loud explosion. He grabbed his camera and ran out of the motel towards the crash.[7]

> By the time I reached the rear of the motel there was a second explosion, not so severe as the first one, and a very large column of smoke was rising from the pine woods just on the west side of I-26. I snapped a picture of the column of smoke and then ran across a vacant field, climbed a fence, crossed both the lanes of I-26 and climbed another fence. Just as I was climbing the second fence a much smaller explosion occurred as I stopped to take a snap of the burning plane. By this time I could see the outline of the plane quite distinctly and could make out the letters "PI" on part of the fuselage. It was then that I realized it was a Piedmont passenger plane.[8]

It did not take long for LaViers to determine that there were no survivors. He turned around and walked quickly back to the motel. Noticing

6. "A large ball of smoke and flame"

Photograph taken by a witness just seconds after both planes crashed on Camp Pinewood (courtesy LaViers family).

that the aircraft had fallen on a cable that connected the Holiday Inn to the phone lines, he stopped a motorist on the interstate and asked him to turn around and drive to the airport to report the accident directly. He made his way back to his room and made himself available to anyone who would need any information on the crash. He also made arrangements to develop the film of the photographs he had taken. He figured there might be some investigators who would need them.[9]

Verne Davis waited for his wife on Ashe Street just off Seventh Avenue. A plant worker at Olin Mathieson Chemical Corporation, the forty-five-year-old glanced northward and noticed the Piedmont jet climbing in the sky. Out of the corner of his eye, he saw a smaller plane approaching the jet at what looked like the same altitude. "Now this is going to be interesting,"[10] he thought to himself. In a statement taken the next day, Davis revealed that he "knew they were going to come close to each other, being quite certain that the small plane would pass over the large jet, which would give the small plane a good scare."[11]

As they got closer, Davis described the scene:

> The small plane pulled up in a steep climb, he seemed to stall for a second. The cockpit and the section forward to the wing of the jet got past the small plane. The small plane struck the wing tip on the left front side of the jet, throwing the small plane into the fuselage somewhere near the engine of the big jet. The little plane completely disintegrated on impact."[12]

Davis saw a large ball of smoke and flame, followed by a loud boom. He watched in revulsion as debris began falling from the large plane as it began, "veering to the right, then to the left, and again to the right; then it passed from my view."[13] The stricken jet disappeared from his view behind some trees and he waited for the explosion he knew would come from the plane's impact with the ground. He only had to wait a few seconds.

Sixty-year-old Joel Dermid had worked his fields for years. A little after noon, he stood in his field looking up at the Piedmont jet in a banking left turn heading east. Looking east, he noticed a smaller twin-engine plane coming toward the jet. He knew they were going to hit each other. In a statement given to the National Transportation Safety Board, he wrote,

> The small plane hit the jet about thirty or thirty-five feet back from where the pilot is. I saw a big ball of fire, a puff of smoke, and a loud noise. I never did see the small plane after the crash but there were all kinds of pieces falling, big pieces and little pieces. When it hit the ground, a cloud of black smoke billowed way up in the sky....[14]

Dermid also commented on the weather. "The weather was clear and the sun was shining and I didn't have any trouble seeing either plane," he continued. "I could see the small plane for ten to twelve seconds before they hit."[15]

Sitting on the front steps of her home at 711 Oak Terrace, Mrs. Robert C. Wilson observed the Piedmont jet.

> I observed what I knew to be a Piedmont jetliner on its departure climb from Asheville airport. I focused my complete attention on this aircraft.... I would estimate that about ten to fifteen seconds after I first saw the jet it passed through a cloud not of clearly defined edges. It had been suggested to me that the plane went behind a cloud; that is, the cloud between myself and the jet.[16]

She continued,

6. "A large ball of smoke and flame"

Rescue and recovery vehicles are parked adjacent to Camp Pinewood on Interstate 26 (courtesy Mac Brackett).

> At one point I had a brief glimpse of the plane in a thin section near the center of this cloud.... After leaving cloud cover, the plane traveled on for a short distance well out of the cloud. At this point the exhaust usually seen behind this type of aircraft was completely and definitely visible as it had been from my first sighting. It was at this time a ball of fire seemed to erupt from the rear of the tail section. I watched it disappear. I started indoors to call the authorities.[17]

As for the weather, Mrs. Wilson wrote, "The sky was sunny but hazy with clouds. There were no other clouds directly in the path flight that I observed the jet follow or between my line of vision and the jet."[18]

As Roy Beddingfield got out of his car in the Berkeley Mills parking lot, he heard "what sounded like a jet plane passing the sound barrier."[19] He saw a "passenger jet slipped over on its side with the left wing down and falling. The aircraft was obviously out of control and the engine was not running as it went down beyond the tree line. Approximately three minutes later debris was seen and started to drift down in the nearby vicinity."[20]

The counselors at Camp Pinewood had another busy day on their schedule. The perfect weather permitted all sorts of outdoor activities for

Onlookers gaze at the wreckage alongside Interstate 26 (courtesy LaViers family).

the day: horseback riding around the stables, canoeing on the lake, and volleyball on the perimeter of the site near the archery range located at the rear of the camp near the trash dump. Sitting in a chair at the stables watching some children, William Murray looked up and saw the Piedmont jet and then directed his gaze back to the children. He then heard a noise he thought was a sonic boom. He looked back up at the Boeing 727 and watched it lose altitude. He did not need to make any trajectory predictions on where the aircraft was coming down. He knew it would land on Camp Pinewood and screamed at his charges to take cover inside the stables.[21]

Head camp counselor Bob Kaufman watched with satisfaction at the great game being played on the volleyball court by the archery range. He stood to the side of the net as a quasi-referee, willing to make a call only if an argument broke out. The server threw the volleyball into the air and

6. "A large ball of smoke and flame" 57

Unidentified recovery workers combing through the wreckage of the aircraft at Camp Pinewood (courtesy Mac Brackett).

brought his fist back to punch it over the net; instead, he let the ball drop to the ground in front of him and started screaming about an airplane. Kaufman heard a whistling noise and quickly looked up. He saw a jetliner on its back heading straight for the volleyball field. Instinctively, he pointed toward the lake and yelled, "Run for the lake, boys!" as both teams took off toward the water. He stopped for a split second making sure the boys ran and immediately realized two of them would never make it out of the crash zone. They had been watching the game by the trees, sidelined by ankle injuries. Kaufman grabbed both of them, one in each hand, and threw them to the ground, falling on top of them to shield them as well as he could. He held the boys tightly, tensed his muscles, and waited to die.[22]

At the lake, Kitty Powers, head girls' counselor, from Miami, looked up in the sky and stared wide-eyed at an upside-down jet falling from the sky. Looking around the shore, she saw nothing but sand and knew there was no time to dig a hole and hide. She quickly remembered the canoes and, with great relief, saw they were still on shore. She ordered the girls under the canoes and followed them, waiting for the terrible explosion and destruction she knew had to come.[23]

7

"We think it might be a big one"

Inside the Boeing 727, the rush of outside air through the hole in the left side of the cabin probably erased much of the agony for the passengers. Whirling around the seats, the force and deafening sound of the air would have created complete confusion among those still conscious, blocking out sensory perception of reality.

Those closest to the explosion, in the forward galley just behind the cockpit on the left side, absorbed most of the shock from the fire and perished instantly, their fatal wounds caused by the flame and debris that burst into the cabin with the force of an artillery blast. As the wind tore open the 727's fatal wound, the protection the fuselage offered grew smaller. The gash broke through the bulkheads as the wind peeled back the aircraft's skin along the left side and over the top.

The collision ripped twenty passengers from their seats and tossed them outside the aircraft into a freefall toward the ground.[1] Some passengers fell whole; others fell in pieces, their bodies shredded by the jagged metal frame exposed when the aircraft's skin peeled away. Just before the plane hit the ground, most aboard were likely already dead or unconscious. Fifty feet above the earth, the aircraft crashed through the trees of Camp Pinewood. Less than a second later, the plane smashed into the earth, creating a gaping, burning hole.

The rumble of the impact reached Bob Kaufman's ears first. He clutched his charges tightly beneath him, expecting debris would rain down on them, but felt nothing. Kitty Powers felt the sand quake beneath her as she, too, waited for debris to come raining down on the canoes. None did. She raised the canoe a bit and saw smoke rising in the distance but no debris anywhere near her. She got the girls out from under the canoes and

7. "We think it might be a big one"

rushed them to the far end of the camp opposite the crash. Sensing that was the safest place, other counselors gathered their charges and headed there too.[2]

Kaufman knew exactly why he and the campers were still alive. The plane had dropped practically straight down. He explained that if it had come in at an angle, the explosion would have spewed debris from all sides, killing anything alive in its way. Since the aircraft came in straight down, the explosion blew straight up, sparing those at Camp Pinewood certain death. Kaufman got up and, dragging his two campers with him, ran to the other side of the camp as fast as he could. Behind him, the demolished, burning passenger jet unleashed a large black cloud into the sky.[3]

The Boeing 727 crashed in an inverted position on a heading of 340 degrees, facing completely the opposite direction it had been flying in before the collision. The debris field started approximately two miles beyond Camp Pinewood and was half a mile wide. The half-mile farthest from the crash site held most of the pieces of the Cessna. The only missing part not found there was the left engine, which was later found imbedded inside the jet. The force of the 727's impact with the ground was so severe that the left wing of the aircraft was found on the right side of the craft and the right wing on the left side. Several major parts of the forward fuselage of the 727 were found along the beginning of the debris path, indicating that the cockpit probably broke up while still plunging to earth.[4]

Accident investigators wrote that "the impact area measured 150 feet by 200 feet with no evidence of horizontal or lateral displacement after ground impact,"[5] which, as Bob Kaufman stated, saved the lives of those on the ground.

Along the wreckage path, stunned residents of Hendersonville walked among the dead, as well as the smoking debris. A house along Orr's Camp road leading to Camp Pinewood had its roof pierced by a dead body falling from one the planes. Other bodies, or parts, lay strewn along the ground from the Blue Ridge School for Boys on U.S. 64 all the way to the wreckage site. Twisted metal lay on the ground, and dodging it, some residents started towards Camp Pinewood to see if there was anything they could do. Others yelled at their children to get inside or, if they stayed outside, not to touch anything. Some called the police, but others found their phones dead due to the aircraft cutting power lines during the crash.

Phone calls to the authorities were not needed. Many of the employees of the General Electric plant, as well as the mills in the area, were also volunteer firemen. With the crash happening at the beginning of their lunch break, many saw the crash and sped towards their fire stations to respond.

Within a few minutes, all available rescue workers in every town and community within driving distance from Hendersonville to Asheville sped towards the collision. In all of those places, a cacophony of sirens rose in the air warning everyone in the area that something really bad had just happened.

In the control tower back at the Asheville airport, Watkins, Brown, and Welch continued to direct traffic, unaware of the horror smoldering on the grounds of Camp Pinewood nine miles to the south. They continued directing aircraft throughout the area and watched as Paul Snell landed Piedmont Flight 1022 on runway 16. They cleared a Beechcraft single-engine plane, N3726Q, to take off and watched as it lifted into the sky.

Flying his Mooney aircraft at 8,500 feet over the Asheville VOR, J.J. Barnes, a thirty-nine-year-old Fayetteville, North Carolina, businessman, scanned the sky for traffic and marveled at the beauty of the mountains below him. He had two other associates with him, and all three were headed for Nashville, Tennessee, on a business trip. As he glanced out the left side of the cockpit window, he saw a small puff of black smoke and immediately noticed what appeared to be the right wingtip tearing away from a jet aircraft approximately a mile away from his position. He watched in horror as the aircraft went into a flat spin and began falling in that spin towards the ground. He turned his aircraft to the left and headed for the crash site.[6]

"Asheville tower, four one five one whiskey," Barnes barked into the radio, trying to keep himself calm.

Evan Brown answered on the local ground control radio frequency. "Four one five one Whiskey, Asheville tower,"[7] he responded.

Barnes took a deep breath. Maneuvering around the black smoke and still-falling debris, he quickly took in the scene and reported it to the tower. "We're just crossed over the city of Asheville and we saw an airplane crash there beside the super highway,"[8] he said, still analyzing the scene below him. The time was 12:04:12.

It took eight seconds for Brown to respond. "Four one five one whiskey, can you circle the site?"

"It's right over a big cloverleaf west of Asheville. We did circle it and, uh, I don't know what the status was."[9]

By this time, Welch and Watkins had heard Barnes's call. All three in the tower looked at each in silence. Immediately, Watkins started looking for Addison's Cessna 310. "3121 Sugar, Asheville."[10]

Keeping his eye on other traffic at the airport, Brown directed Snell's Martin 404 to the ramp, while his mind raced to figure out what had just happened. "Piedmont ten twenty-two, cleared to the ramp,"[11] Brown said, watching the Piedmont Martin 404 turn off the taxiway.

Trying to keep his cool, Watkins repeated his call again. "Cessna 3121 Sugar Asheville,"[12] he said, placing an emphasis on Asheville in an attempt to let Addison know this was serious. The crashed plane could not be Addison's, though, Watkins thought. This 4151 Whiskey reported a crash to the south, but he had sent the Cessna to the north. He quickly reviewed his clearance with Addison.

"Ten twenty-two,"[13] Snell acknowledged Brown's command. By now, he too had heard the news on the local controller radio.

"Four one five one whiskey, could you, uh, determine what type aircraft?"[14]

By now, Barnes had flown over the crash site, made a few observations and steered his aircraft back towards Nashville. "Uh, Asheville tower, this is 4151 whiskey; looks to be a B, ah, B forty, uh, five, or a B fifty-two. It looks like the wing tip blew off of it."[15]

Although unspoken yet in the tower, all three controllers knew which aircraft lay in smoking pieces on the ground. Still, it had to be confirmed. Steeling himself, Brown asked the vital question: "Was it a jet?"[16]

"Affirm. Do you not see a [sic] large black smoke rising around?"[17]

Brown did not answer. Instead, he instinctively started looking for Flight 22. "Piedmont 22, Asheville,"[18] he queried into the radio. Hearing no answer, he waited seven seconds and tried again. "Piedmont 22, Asheville tower."[19] He waited in silence and did not hear a response. A full twenty seconds passed before Brown contacted Barnes. "4151 whiskey. Are you still in the vicinity of the crash?"[20]

The interphone line rang. "Atlanta,"[21] the caller said.

Watkins had no time for this. He had to find Addison's Cessna 310 and—it looked like—the Piedmont jet before he spoke to Atlanta. He knew they were calling for each aircraft's status and at this second he did not have an answer for them. He had an idea about it, but the thought made him ill. "Asheville, I'll call you back,"[22] Watkins said, hanging up the phone abruptly.

"Okay,"[23] the Atlanta controller said to nobody.

"3121 Sugar, Asheville approach," Watkins spoke into the radio. Ten seconds later he repeated it. "Cessna 3121 Sugar, Asheville approach."[24]

From his aircraft, Barnes finally replied, "Affirm. We're about five miles, it looks like, south of it and the smoke off the crash is up at my altitude and I'm at nine thousand. There are small aircraft flying around it. We see them below."[25]

Brown could not believe it. "Say that, say that last again?"[26]

"There are small aircraft flying around the scene, ah, real low, near the scene,"[27] he repeated.

"Roger, thank you,"[28] Brown responded. That's all he needed right now—a bunch of aircraft flying together in a small space, looking at the ground and not at each other. He wondered who they were.

Sitting in the office of the small Meyer Airport on the outskirts of Hendersonville, three men—twenty-one-year-old Clifford Port, a flight instructor for Meyer Flying Service; retired Air Force Colonel N.H. Van Sicklin, a pilot and sales representative at the airport; and Oscar Meyer, the airport operator—all sat down to lunch when Ralph Bishop, an airport mechanic, came running in, pointing out the window and telling them that a plane had just crashed. The three men, followed by Bishop, ran outside and saw a sinister black cloud rising up through the sky northeast of their location.[29]

Sensing a disaster, they all ran toward their parked airplanes. Van Sicklin hopped in his Cessna 150 and Bishop climbed into the passenger seat. Meyer took off in his Cessna followed by Port in his. All three planes took off and raced towards the crash site to try to figure out what happened and what aircraft was involved.

"Asheville tower, this is 15 Juliet," Van Sicklin said from his aircraft.

"Fifteen Juliet, Asheville tower," Brown replied.

"Did you have this aircraft, ah, inbound to Asheville that just crashed near the Holiday Inn?" Van Sicklin asked.

"Fifteen Juliet, I haven't been able to determine which one it is. Now, are you the light aircraft that's circling the scene of the crash?"

"Yeah, I'm one of the three,"[30] Van Sicklin replied.

The phone in the tower rang again and Watkins answered. It was the United Air Lines representative at the airport. "Did someone report to you a downed airplane?"

"Yeah."

"Okay. You have all you need," the UAL representative said in a matter-of-fact tone.

Watkins snapped, "No, we don't have all we need. You got anything?"

"We have the person who just called us; his name and phone number."

"I don't need that," an exasperated Watkins barked. "Is he, right now, does he know anything about what kind of airplanes, how many of them?"[31] This was the first time anyone in the tower verbalized what they had been thinking: two planes may have been involved in this crash.

7. "We think it might be a big one"

"He said that it appeared to be a C-141. He's a sergeant in service assuming that he…."

Watkins cut him off. "OK. Hang onto that, Bob. I'll talk to you later."³² Watkins hung up.

Two planes, he thought. Please, God, not here, not on this shift. He struggled to keep his cool. "Can you determine what type of aircraft?"³³ he asked Van Sicklin.

Van Sicklin flew lower toward the trees, getting as close as he could without becoming blinded by the smoke and falling debris. "No. It's pretty well busted up. The tail looks like Howard Loadstar or something but I thought it sounded like a jet when it went in," he replied. He paused a few seconds then continued. "It's green and white I believe. Green and white or blue and white, I can't see the number; almost, but not quite."³⁴

Brown knew he had to ask. He had to verbalize to everyone on the frequency what he, Welch, and Watkins already knew. "Roger, does it appear to be one aircraft or could it have been two?"³⁵ There, he had said it. He just suggested over the air that two planes under the control of his tower may have collided.

"I can only see one," Van Sicklin replied.

"Roger,"³⁶ Brown replied. He knew that if a Boeing 727 collided with a Cessna 310, the Cessna would probably disintegrate.

Listening in disbelief in his cockpit a few hundred yards from the control tower, Paul Snell hurried his passengers off his aircraft. "What position is this, tower, from here?"³⁷ he asked over the radio. He knew the direction his company's Boeing 727 had taken off in, had spoken to the crew just a few minutes before. When he heard the word "jet," he knew which plane was involved.

Van Sicklin heard Snell over his radio, as both were using the same frequency. "Well, it [*sic*] right across, just west of twenty-six right on the edge of it and directly across from the new Holiday Inn here in Hendersonville."

"Roger,"³⁸ Brown responded. He appreciated Van Sicklin's detailed description.

The phone rang and Watkins picked it up. It was Atlanta calling again. The controller sounded agitated. "Piedmont 22, can you switch him to me yet?"³⁹

Watkins could not hide the news any longer. He knew Atlanta had figured out something was wrong because they had not received an update from him and they could not find the Boeing 727 on their radar or radio. Watkins unburdened himself. "Um, afraid I can't right now. He's unreported and this three-ten that was inbound was cleared over to Asheville beacon and he's also unreported," he admitted. He got himself together and continued. "And, ah, we have a public report of a crash around Hendersonville."[40]

It took almost eleven seconds for the Atlanta controller to get over the shock of what he had just heard. "OK," he said.

"So, I'll let you know," Watkins added, but the Atlanta controller had already hung up. Watkins grabbed the radio. "Cessna 3121 Sugar. Asheville approach control."[41] He knew it was futile and finally admitted to himself that both planes had probably collided south of the airport, but he could not understand what the Cessna was doing down there.

Snell asked the tower if he could take a look. "Tower, uh, we'd like to go down that way as soon as these kids get off. We'll give you a pretty good report when we get down there,"[42] he added. Snell knew he had to see for himself, the tower, and his employer what had just happened.

"Roger," Brown said, granting his permission.

Snell couldn't help himself any more. "I'm afraid to look," he added for all to hear.

"Yes,"[43] Brown admitted. He too was afraid to look, but he had a better idea than Paul Snell of what had just happened.

Harold Roberts, the FAA tower chief in Asheville, bounded up the steps to the control room. Having monitored the radio, he already figured out what must have happened. He did not need an explanation.

Roberts snapped up the ringing telephone. "Control tower!"

"Who is this?"

"Roberts."

"Roberts?"

"Yeah," he replied, recognizing the voice on the phone.

"What you got down over yonder?"[44] Lacy Griffin asked. Griffin ran the Asheville Flying Service, and his name was synonymous with Western North Carolina Aviation. He owned the flying service at the old airport and moved it here when the new airport was built. He was one of the few guys Roberts would allow to ask a question right now.

"I don't know, Lace. We think it might be a big one."

"Big one?"[45] Griffin asked incredulously. Big ones just started flying here to the mountains, he thought. He hoped Roberts was wrong.

"Yeah."

"Oh, Lord,"[46] Griffin replied, knowing the horrible implications of a major air disaster.

"Airplane reported a wing coming off," Roberts continued. He had to tell someone; it was like confession. "Down there about Holiday Inn, best we can figure."⁴⁷

For a man who had flown propeller aircraft his whole life, Griffin automatically thought of a DC-4 or an L-1049 Connie. Jet aircraft were as new to him as they were to the mountains. It was not until this moment that he realized what Roberts was saying. He took a deep breath and asked. "Big airplane wasn't a jet, was it?"

"Could be, I don't know," Roberts responded quietly.

Griffin shook his head. "Okay, thank you."⁴⁸ As a seasoned professional, he knew that he did not need to be flying around the crash site, so he walked back to his office in case anyone needed him or his advice. He kept the radio on.

Other inquiries about the jet on the ground came in while the controllers tried to keep traffic flowing at the airport. The struggle of trying to find out the details of the crash, keep the traffic flowing, answer the phones, while at the same time review every procedure that took place over the last half hour—because they knew there would be questions, a lot of questions, about their role in this—had started to take its toll. Roberts was starting to ask questions about what happened, so Brown contacted the pilot of 4151W again to get a report of what he saw.

Roberts grabbed the radio from Brown and listened intently as Barnes repeated his tale of horror. "We saw this aircraft in the air that crashed like, ah, west of the Holiday Inn. We saw the right wing tip depart from the aircraft and it looked like an explosion in the air. Tremendous amount of black smoke at the time that the wing tip came off and it went right on to the ground."

"Ah, 51 whiskey, was that a jet?"

"Affirmative," Barnes replied. "It looked like a jet here, Asheville tower. We're departing from the scene, heading on our course to Nashville,"⁴⁹ Barnes said. He had seen enough here and wanted to leave the grim sight and get on to Tennessee.

Harold Roberts now knew what aircraft was lying on the ground across the street from the Holiday Inn. "51 whiskey, roger. Thank you very much."⁵⁰

The phone rang again and this time Watkins answered. It was Atlanta control calling for another update on Asheville's aircraft. This was not the same controller as earlier; he hadn't heard about the two missing aircraft. The first call had taken place seven minutes earlier. After giving him an update on the aircraft still in the air, Watkins turned to the two missing aircraft. "We've got 3121 Sierra inbound. I gave you out of seven on him."

"Three-one two-one Sierra?"

"You've taken him down already,"[51] Watkins said, figuring out they must have taken that aircraft's strip off their wall when they handed it to Asheville.

"Yes,"[52] Atlanta control replied matter-of-factly.

Watkins struggled with the next group of words. "He was out of seven and he's unreported."

"He's unreported?" the controller asked incredulously.

"Right and we've got this Piedmont twenty-two. You never did talk to him, did you?"

"No, we didn't."

"He's unreported,"[53] Watkins told him.

Watkins waited for the pause to end on the other end of the line.

"Okay," Atlanta replied. "As soon as you find out something, advise me please."

"Okay."[54]

Sensing that the Asheville tower needed more information, Clifford Port described the scene above the dead jet. "There's still a lot of debris in the air. Still falling and it's quite a bit scattered over a wide area so that report of that wing coming off may be about right," he said, assessing the situation.

"Ah, thank you,"[55] Roberts responded.

Now the controllers had to deal with Paul Snell. "Piedmont taxiing out," Snell said into the radio.

"Piedmont, are you going out as 7430?" the tower asked, verifying his new flight number.

"Yessir, 7430."[56]

Not sure where he was going, Snell asked for directions. "That's just down the highway here; runs by the airport, doesn't it?"

"Say that again," Brown repeated.

"I-26 is the new one here by the airport, isn't it?"

"Affirmative. That is the interstate that goes down southwest and the scene of the crash is reported near Holiday Inn at the cloverleaf,"[57] Brown said, repeating the latest information on the location of the wreckage.

"We're ready to go," Snell said a minute later, and then the tower cleared him for takeoff. "Here we go,"[58] Snell said into the radio, with a tone that implied he was getting ready to start a grim mission.

Van Sicklin called the tower and gave them an update on the crash site. "The debris is oriented north, correction, southwest from the scene of the crash and there are pieces all over in the trees and on the ground."[59]

Now for the first time, fifteen minutes after its first notification,

Asheville tower realized the danger of having numerous planes flying around the crash site.

Brown knew he needed to warn all the aircraft by the Holiday Inn about each other. "More traffic is converging on the area. A Cessna 310 just departed from the Asheville airport going down that way and a 404 just off now, he'll be going that—to look at it."[60]

"We'll be watching for him and this debris,"[61] Van Sicklin replied. He knew his job was about done. He noticed that Meyer's aircraft as well as Port's had already started back to the Hendersonville airport. "This must have happened pretty high, because debris is floating down from high altitude in small pieces."[62]

"Roger, understand debris is still floating down in small pieces all over the place?"

"That's roger,"[63] Van Sicklin replied, turning his aircraft away from the site to let the others have a view. "It's floating down in small pieces all over the place."[64] There was nothing more to see, and if there was he would gladly let others get their view. One can only watch a horror movie for so long.

8

"Deliberate, continuous leaks"

Within half an hour of the accident, news of the crash reached the National Transportation Safety Board (NTSB) in Washington, D.C. Since being formed just a few months before by the Department of Transportation Act of 1966, the board had been extremely busy. On June 6, 1967, the NTSB had held three days of public hearings in Dayton, Ohio, on the midair collision that occurred in March over Urbana, Ohio, killing twenty-six people. Investigators continued to gather evidence in the crash of a BAC-111 near Blossburg, Pennsylvania, on June 23, 1967. Thirty-four people perished in that crash. The board had already penciled in the date of August 2, 1967, to hold a public hearing on the crash of a Lake Central Airlines accident near Marseilles, Ohio. That accident on March 5, 1967, killed thirty-eight passengers and crew. Even as the *Manhattan Pacemaker* lay smoldering at Camp Pinewood, a public hearing was being held in New Orleans, Louisiana, on a Delta Airlines crash that killed nineteen. None of those crashes, however, compared to the disaster that now faced the NTSB investigators: a midair collision in which eighty-two perished in an instant.

For many years, investigators at the Civil Aeronautics Board (CAB) were responsible for aircraft accident investigations. Established in 1938, the CAB regulated the nation's airways. The CAB decided which airlines would be rewarded with a route, what that route would pay, and how much profit the airline would make. It did not crush competition; it regulated it. As Whitney Gillilland, a CAB member, stated, the board's functions are "the promotion of air transportation through regulated competition."[1]

Soon after taking office in 1963, President Johnson started toying with the idea of combining all of the governmental departments charged with

8. "Deliberate, continuous leaks"

safety and placing them under one roof. He knew the benefits of having all transportation-related agencies under one large department. Hence, the Department of Transportation (DOT) Act of 1966 was born. Johnson shuffled the FAA under the DOT shell but as an autonomous department. Highway safety, pipeline safety, etc., were all carved away from their predecessor agencies and placed under the DOT umbrella. Johnson stripped the CAB of its investigative functions, moved the entire CAB air safety board to the DOT, and named it the National Transportation Safety Board (NTSB). While it operated under the DOT name, it was in theory an autonomous department with a large degree of independence.[2] The board's success, reputation, and credibility rested on its ability to investigate accidents without interference from outside influences. Congress baptized the board with independence at its birth because it believed the board could not operate successfully any other way.

With the CAB's air safety board now fully entrenched as the NTSB, the board lacked leaders. President Johnson wanted the board set up like the CAB, with five members appointed by him for staggered terms. In order to assure fairness, the members had to be from both political parties. On April 17, 1967, the United States Senate held a hearing on the nominations Johnson sent down from the White House. He assigned Joseph J. O'Connell, Francis H. McAdams, Rear Admiral Louis N. Thayer, Governor John H. Reed, and Oscar M. Laurel to the board and awaited Senate confirmation.

Joseph O'Connell came to the board with a lawyer's background in aviation. He formerly headed the CAB, having been appointed to that position by President Harry Truman. He served many clients over the years, including Japan Air Lines and Beech-Nut Life Savers, and he was also chairman of the board for Lake Central Airlines, the very same airline that lost a plane in Marseilles, Ohio, just a few months before.[3]

Attorney Francis H. McAdams, CAB assistant to Whitney Gillilland for the last seven years, came before the Senate; Gillilland described him as a man who does not make mistakes. At McAdams' Senate confirmation, Gillilland stated, "I have never known him to fail or make a mistake, which is a pretty strong statement, but it happens to be accurate in his instance."[4] Gillilland further gushed, "There has been no case that the board decided in this field during that length of time that did not meet his very careful scrutiny and study.... I think perhaps that he has had more to do with the final form that the safety decisions have taken than anyone else at that level."[5] That included clearing the Boeing 727 after its rash of crashes in 1965.

Coast Guard Admiral Louis Thayer was at the end of a distinguished career with the Coast Guard and Merchant Marine. He had safely escorted over five thousand refugees from Cuba and directed many rescue operations. He also led the investigation into the fire that broke out on the *Yarmouth Castle* luxury liner. His expertise included radar plotting and anti-submarine warfare.[6]

John Reed became the only Republican appointed to the board. A very successful potato farmer, he was elected to the Maine House of Representatives and then the Senate. His colleagues voted him president of the Senate, a position that placed him next in the line of succession to be governor. When Governor Clinton Clausen died in office, Reed suddenly found himself governor. After winning a partial term as governor in his own right in 1960, he was reelected in 1962. In 1966, however, Maine voters turned Reed out and elected Kenneth Curtis as governor.[7] Even though he was a Republican, Reed was a staunch supporter of Democratic President Lyndon Johnson.

Lyndon Johnson found his fifth member in his home state of Texas. Oscar Laurel was a former investigator for the district attorney's office in Laredo, Texas. Voters in the 80th District in Texas elected him as their state representative for two terms. In 1964, he became a state coordinator of the Viva Johnson clubs of Texas, helping run the Texas presidential campaign for Johnson.[8] The president rewarded him with a spot at the NTSB.

O'Connell became the chairman of the board, and at the confirmation hearing he spoke for the board as a whole. In discussing the responsibilities of the new board, O'Connell supported the policy of the board's independence. "We will exercise the responsibility completely independent of the secretary of the Department of Transportation,"[9] he told the confirmation panel.

O'Connell and the Senate panel engaged in fantasy if they believed that the board would remain independent from outside influences. Each member of the board owed his position to the president and would be reappointed, or not, depending on his pleasure. The transportation secretary also controlled the purse strings of the board. Their upcoming budget for 1968 was made up by analysts within the DOT. The administrative duties of the board were also handled by DOT personnel.[10] The board members and their employees at NTSB were held accountable to those who had the power of appointment and the power of the purse.

Conflict-of-interest issues also arise when a government agency such as the NTSB investigates another government agency such as the FAA, especially when both are under the umbrella of a larger department, in this case the DOT. In turn, the executive branch of the government controlled the DOT, and Lyndon Johnson, facing the pressure of an unstable economy, controlled the executive branch. Cuts had to be made in programs lest the war in Vietnam and his Great Society programs go unfunded. Money became scarce, and the NTSB was well aware that. As of April 30, 1967, there were 433 suits outstanding against the government for claims totaling $203 million as a result of aviation accidents.[11] As it stood at that moment, there were 203 million reasons why the NTSB could not remain independent in this, or any, case where it investigated another government agency.

Even worse, investigators of a major aviation accident were, and still are, dependent on other organizations, those with a financial interest in the crash, to help them determine the probable cause of an accident. This practice is called the party system and is still used by the NTSB today.[12] Because of the small size of the NTSB, the board is not adequately staffed to investigate all aspects of an aviation accident without outside help. For example, if a Boeing aircraft crashed, Boeing employees help the NTSB investigate the accident. Pratt and Whitney employees are called in if that company's engines are involved. The Air Line Pilots Association (ALPA) is generally always a party to the investigation if the involved aircraft was conducting a commercial flight. Representatives of the involved airline are also asked to take a seat at the investigative table. All of these groups have a financial interest in the outcome of the case and generally became defendants if a case went to trial. The party system allows future defendants to investigate the crash. Under the party system, the conflicts of interest are so obvious that it is doubtful that a thoroughly complete and unbiased investigation could ever be conducted by the NTSB.

Except for family members of those lost in the crash, the news hit the Piedmont offices in Winston-Salem the hardest. Immediately, top airline officials raced to their private planes and took off west towards Asheville and the horror that awaited them. Harold "Zeke" Kimball Saunders, vice president of operations for Piedmont, arrived a few hours after the crash. One of the first employees of Piedmont, Zeke had been a military pilot during World War II, and Piedmont hired him at the end of it. He answered only to Tom Davis, founder and president of Piedmont Airlines.[13]

No one knew better than Saunders that because of his position as vice president of operations, his actions, as well as the actions of the Piedmont pilots, would be scrutinized thoroughly. Investigators would look at pilot training, flight time, Piedmont policies and procedures, maintenance records, etc. Anything that had to do with the operation of a Piedmont aircraft fell under Zeke's jurisdiction. This crash had placed him under a microscope, and he knew that any mistake by the pilots, any action that violated company policy, could place Piedmont in a position to pay out millions of dollars in liability.

Zeke also knew, however, that Piedmont employees would be a part of that investigative team and that Tom Davis gave him the power to make those appointments. It was obvious to Zeke, as it would be to any vice president of any airline involved in an accident of this magnitude, one never experienced on this scale by Piedmont, that those appointments would be vitally important. He would fill them with those he trusted and those who held the interests of Piedmont Airlines paramount during the investigation. He would be derelict in his duties if he acted any other way.

After arriving in Hendersonville, Zeke surveyed the crash scene and immediately took charge of Piedmont's response. He booked motel rooms for Piedmont personnel near the site and made a list of those he wanted to serve as Piedmont's representatives to the NTSB investigation.[14] He then waited and wondered if his brother, Tom, would come.

Zeke met his brother at the accident site, and the men exchanged greetings. However, Zeke wrote that "upon learning the next day that my brother, Thomas Saunders, was the National Transportation Safety Board's investigator in charge [IIC] of this accident investigation, I turned over my responsibilities regarding the accident investigation to other Piedmont employees. From that point forward, I did not participate in the NTSB's investigation."[15]

In its first major accident investigation, the NTSB ignored its vows of impartiality to Congress, killed any credibility it had with the public, and exposed itself to charges of conflicts of interest by its assignment of Tom Saunders, brother of Zeke Saunders, as IIC of this crash.

As the day dragged on, investigators roped off the scene of the crash and began their work. Tom Saunders split his investigative team into several groups. Each group had the responsibility of investigating different aspects of the crash, including analyzing the flight data recorder and cockpit voice recorder from Piedmont 22, interviewing witnesses, reviewing air traffic control procedures, and examining the engines on both aircraft, the weather, the different systems on each airplane, human factors, structures, maintenance, and operations.

As IIC, Tom Saunders inherited the responsibility to appoint parties to the investigation. In this particular crash, the choices were easy. He appointed the FAA, Air Line Pilots Association (ALPA), Air Transport Association (ATA), Air Traffic Control Association (ATCA), Aircraft Own-

8. "Deliberate, continuous leaks" 73

ers and Pilots Association (AOPA), Pratt and Whitney, Cessna Aircraft Company, the Boeing Aircraft Company, and, of course, Piedmont Airlines. Each party appointed its representatives to assist in the investigation. In the case of Piedmont Airlines, Zeke Saunders made the appointments. Incredibly, many of the individuals assigned by ALPA as its representatives were the ALPA members who worked for Piedmont Airlines. Piedmont employees now made up two groups investigating the crash: ALPA and Piedmont Airlines itself.

Out of the seven members of the ATC group, two were Piedmont Airlines employees; one represented Piedmont in the investigation, and the other represented ALPA.[16] On the power plant group, one out of four worked for Piedmont.[17] Two-thirds of the members of the weather group worked for Piedmont: one assigned to represent Piedmont, the other representing ALPA.[18] Over half of the members of the systems group worked for Piedmont: two represented the airline; one represented ALPA.[19] One-third of the members of the flight recorder group worked for Piedmont.[20] Half of the four maintenance group members worked for Piedmont: one representing the airline, the other representing ALPA.[21] Also, over half the members of the operations group were employees of Piedmont: two represented the airline, and two represented ALPA.[22] Other groups were set up the same and were weighted heavily in favor of Piedmont Airlines as compared to any other party in the investigation, especially those interests representing the Cessna and its occupants.

All Piedmont employees assigned to investigative groups knew the consequences of their discussions, analysis, and conclusions regarding the accident. The financial ramifications of a detrimental finding against Piedmont could be catastrophic to the airline. The traveling public quickly becomes wary of airlines that are involved in crashes, especially an airline that is at fault. The outcome of this investigation could bankrupt Piedmont and be the end of the airline as well as the end of the jobs for all the Piedmont employees on the committees.

The creators of the party system set it up to augment the investigative process and naively assumed the groups would work in a vacuum. While rules were in place to restrict the flow of information back to corporate headquarters, enforcement of such rules was not realistic. Nothing could prevent an airline executive from asking questions regarding an investigation of an employee who was assigned as a party member to that investigation.

As the afternoon began to fade and the smoke from the crash dissipated, the investigation started to take shape. Tom Saunders appointed the parties, including Piedmont Airlines, while his brother Zeke appointed the party members from Piedmont Airlines. John Reed accompanied Tom Saunders to the crash site to act as the conduit between the press and the board. Reed scheduled a press conference for early that evening. Investigators fanned out across western North Carolina to start interviewing witnesses, especially the three men in the airport tower at the time of the crash.

A different group of investigators spread out to interview those having anything to do with the small plane. FBI agents showed up at the house of Bob Anderson in Springfield, Missouri, peppering his shocked widow with questions about his politics, social habits, friends or enemies, etc.[23] They drove north of Charlotte to interview the men that the Cessna crew spent time with the day before and the morning of the accident. Already, the Cessna crew had been cast as villains. While the NTSB dealt with Piedmont personnel, the FBI dealt with the Cessna's families.

Board member John Reed held his press conference that evening. The NTSB set up its headquarters at the Holiday Inn, right across the street from the crash site. Droning on, Reed explained the responsibilities of the NTSB during an investigation, introduced various members of his investigative team, and set forth what they wanted to accomplish.

Reed did not go into much detail since, as a potato farmer, he did not know much about aircraft investigations. He cautioned against attributing blame for the crash and did not comment on who he thought was at fault at this point, but he added that "no aspect of the crash will be overlooked."[24] If the reporters at the press conference wanted to write something salacious for tomorrow's papers, they were listening to the wrong person.

Harold Roberts was the right person. He went into protective mode immediately after the accident. He knew his responsibility lay with his men in the tower. Relieved of duty late in the afternoon, he and other FAA personnel met to try to figure out what caused the collision. Watkins remained adamant that he had given the Cessna instructions to fly north of the airport, not south, and that Addison must have ignored his command. It was good enough for Roberts and the FAA. When asked after the meeting by a reporter whether they knew what had caused the crash, Roberts stated, "The small plane was about twelve miles south of where it should have been."[25]

Asheville Airport manager Ken Dacy also quickly placed the blame on the Cessna's crew. He stated that Addison "should have been north of the airport, where we have an H marker by the Vanderbilt mansion, but he was south of the field about twelve to fifteen miles from where he should have been."[26] It obviously did not occur to John Reed, or anyone else at the NTSB, to stop these unauthorized leaks.

The verdict had been delivered before the trial was held. The party with the least amount of financial resources to mount a defense, the late David Addison, was quickly blamed for the crash. Now the other parties, with almost unlimited financial resources behind them, would gather all the evidence in a measure of self-preservation and present that evidence in a way to clear themselves and blame Addison.

The crash of Piedmont 22 shocked officials in Washington. From the White House to the Pentagon, through the Mall and Capitol Hill, stunned officials could not believe that two planes had collided again and that this time one of their own had died.

President Johnson issued a statement that day. It read:

THE AIR CRASH in North Carolina has made this a tragic afternoon for many American families. Among those we mourn is a family that many of us in Government know, admire, and will deeply miss. John McNaughton, his wife and youngest son have been taken from us with cruel suddenness. Their loss can only be measured by the emptiness it will leave in American hearts and the pages of American history.

For six and one-half arduous and decisive years, John McNaughton served in the highest councils of our government. His devoted wife served beside him. He was soon to become Secretary of the Navy—and this adds a special poignancy to his death. For it reminds us again of the rare breadth of his abilities, the selflessness of his great energies and talents, the enormity of the loss suffered by all free men who have found, so long, pride and inspiration in John McNaughton's example.[27]

Both Johnson and Secretary of Defense McNamara had to take their friend's death as another setback in their Vietnam policy, a setback they did not need. The killing of American soldiers had risen to over one hundred per day, and Johnson knew the country would not stand that type of body count for long. Citizens protested the war daily, students rioted on college campuses, and the downward spiral continued. Johnson was running out of answers, and John T. McNaughton had taken some of them with him to the ground.

Johnson also measured the death of McNaughton in terms of politics. The cost of the war had increased daily, and the Pentagon had spent tens of millions of dollars to fight it with nothing to show for the money and slaughter. His "Great Society" slowly suffocated every time Johnson ordered his department heads to cut back domestic spending to funnel the money to fight the Vietnamese. He knew how enthusiastic the FAA administrator, William McKee, had been in making those cutbacks. As a former general,

McKee was one of the few men around Johnson whom the president could depend on to do as he was told. Now it had cost Johnson a trusted and close advisor. He must have wondered just how much longer he could take this job.

Secretary of Defense Robert McNamara's world had begun closing in on him about a year before when his aides privately told him that the United States could not win the Vietnam War. Since it was his duty as secretary of defense to relay these thoughts to his commander-in-chief, he did so, and his warnings fell on deaf ears.[28] He knew that the president was in this fight to stay. What shocked and horrified McNamara most about the war was the cost in lives, effort, and money. He had lived and breathed, slept and dreamed about Vietnam since the Gulf of Tonkin incident, but at least up until noon he had John McNaughton around to unburden himself to. Now McNaughton was dead and McNamara could not help wondering whether that death could have been prevented if more money had been spent on air safety than on Vietnam bombing runs. No doubt, he also wondered how much longer he could take this job.

FAA Administrator McKee knew the sum of the whole. He could not believe another midair collision had occurred. He, more than anyone, knew all the secrets regarding the aviation cutbacks of the last few years, cutbacks that were sliced from his budget to satisfy the prosecution of the war.

McKee remembered how proud he was telling the president about all the savings that he had brought about within the FAA. He tried to make a mental list of the items he cut, but it went on and on. He had cut weather station reporting, did not pursue the hiring of air traffic controllers, tried to stave off advancing technology in favor of the current antiquated equipment in air traffic control. He remembered the directives sent, not directly by him, but with his implicit approval, regarding equipment cutbacks and mechanical staffing. He visualized in his head the letter, addressed "Dear Bozo," from the president, and knew he was in trouble.[29] Congress would want answers, and he expected a summons to appear before Congress any time, knowing it would be measured in days rather than weeks. He too must have wondered how long he could take this job.

Joseph O'Connell felt the heat from his job from the moment President Johnson signaled that he wanted him as chairman of the NTSB. There had been no respite from air tragedies since the board was formed in April. Even one of the jets from the airline he had, until recently, run had been involved in a crash. Now a midair collision killing eighty-two people, including one of the president's most trusted advisors, just put more heat

8. "Deliberate, continuous leaks"

on O'Connell. Things had been simpler when President Truman appointed him as CAB chairman in 1948. When a plane crashed then, one would not expect more than fifty people to be killed. Now, in just twenty years, when planes had doubled in capacity and tripled in speed, and airlines were looking for further increases in each, a person could realistically count hundreds of dead under the wrong circumstances.

O'Connell had started his service in government when President Franklin Roosevelt appointed him general counsel for the Treasury Department.[30] He had been in government long enough to know that Congress would demand answers regarding this plane crash, and he had not been on the transportation board long enough to even know what the questions would be. He knew he would not be able to take this job for long.

Kenneth Lyons, president of the National Association of Government Employees (NAGE) and his son-in-law, Stanley Lyman, vice president, FAA Affairs, NAGE, felt like prophets when they received the news of the Piedmont crash. Their union, NAGE, represented approximately eight thousand out of the fourteen thousand air traffic controllers, and they had been telling Congress since last March that the FAA did not have the money, manpower, or motivational leaders to keep the nation's airways safe.[31] What frustrated them most was that no one was listening. They knew more hearings would be called by Congress on air safety and that the same tale would be told and would fall on the same deaf ears.

Something dramatic had to be done to get the attention of Congress and, most importantly, the president. They knew Congress would want to hear from the FAA and the NTSB before NAGE was called. Lyons and Lyman would wait until those aviation officials spoke, as both men were particularly interested in what General McKee had to say. Then they would make their splash. They wondered, after they got through with their testimony, how long any of those guys would keep their jobs.

Halfway across the country, Joe Dando, owner and president of Lanseair, seethed with anger as he continued reading reports in the press blaming his men—David Addison, Ralph Reynolds, and Robert Anderson—for the Piedmont plane crash. He had not even had time to speak to the families of his dead employees before the government started its smear campaign against his firm. On the advice of his attorney, Joe Greene, he decided to send a telegram to John Reed at the NTSB to try to counter these tactics.

Dando stated that he was "disturbed by what appears to be deliberate, continuous leaks and/or official releases from various government officials or employees involved in the investigation of the air tragedy."[32] He was referring to the statements of Harold Roberts, who declared the Cessna was off course, as well as Dacy's remarks.

Dando's telegram to the NTSB continued: "Our representatives have been told that no official information is available but we continue to see what purports to be statements attributable to your officials in the newspapers."[33] He went on about their smear campaign. "The newspapers, quoting these officials, have appeared to assess blame against the Cessna, whereas our investigation indicates quite the contrary. I am told that the hearings are for determination of probable cause and therefore we should not be prejudged by the news media as a result of the actions of your employees.... Please reply."[34]

Other than his attorney, Dando could only count on help from the AOPA, of which Addison was a member. Against him and the AOPA were arrayed the might of the United States government, all the financial muscle of Piedmont Airlines, Boeing, and Pratt & Whitney, and all the other multi-million-dollar corporations that would face liability action if the Boeing 727 or its crew was found to be at fault. Along with Dando, the AOPA protested the government's actions.[35]

Representative Harley Staggers (D-WV) wanted answers quickly. His House of Representatives Committee on Interstate and Foreign Commerce, the umbrella committee for the Subcommittee on Transportation and Aeronautics, had just held hearings in March regarding aviation safety after the Delta and Lake Central crashes. Now, it seemed that airplane crashes had not been reduced but were becoming more common. He had recently come under public criticism from columnist Drew Pearson for holding aviation hearings in private rather than public forums. In his "Washington Merry-Go-Round" column of July 6, 1967, Pearson wrote, "But for some reason best known to himself, the venerable Congressman from West Virginia has chosen to hold airplane accident probes behind a strict veil of secrecy...."[36]

With the small number of deaths involved in those crashes, as well as the relative anonymity of the victims, Staggers probably saw no reason to embarrass the airlines in a public forum any more than they had been by press accounts of the accidents. Or, it could have had something to do with the fact that Joseph O'Connell, recently appointed chairman of the NTSB, had been chairman of the board of Lake Central Airlines from 1955 until last year.[37] But now, with a midair collision killing eighty-two people, including the secretary-designate of the navy and most of his family, there was no way he could keep any upcoming hearings in secret.

Staggers knew the identity of the real culprit. He knew what the president and McNamara were spending in Vietnam. He also knew that McKee had not questioned the president on Vietnam spending. The congressman realized that his committee was responsible for funding aviation safety, and he would be damned if he or his fellow congressmen were going to be blamed for all these aviation deaths. No, he would hold the hearings in public within the next few days. The American people had to know why money was not being spent on aviation safety and where their money was truly going, and if it meant embarrassing the president, he intended to do so. Staggers notified his staff to contact the heads of the FAA, NTSB, and NAGE. He wanted them in front of his committee on Capitol Hill at 10:00 a.m. on Monday, July 24, 1967, for a showdown.[38]

9

"This process is conducted entirely in the open"

Congress had been buzzing over the weekend. Several congressmen audaciously asked how many people had to die before the airways got fixed. Proponents of aircraft segregation—between private and commercial aircraft in the sky—vocalized their position and used the deaths in Hendersonville to bolster their argument. Others proposed building separate airports for private and commercial planes, and at least one had the Boeing 727's dirty past on his mind. All of them, though, had one thread in common: hearings had to be held to find out what was wrong with the nation's skies.[1]

Congressman Staggers summoned the witnesses to his committee's headquarters in the newly completed Rayburn House Office Building just off Capitol Hill. In complying with their summons, Joseph O'Connell and William McKee, along with their respective assistants Bobbie Allen and David Thomas, undoubtedly fell within the gazes of the statues of *Spirit of Justice* and *Majesty of Law* as they entered the Rayburn building. *Spirit of Justice* holds a lamp in her right hand, which symbolizes the light of knowledge and truth; her left arm is wrapped around a child, showing justice tempered by love and protection. *Majesty of Law* is a Zeus-like figure holding a book of federal laws with the seal of the United States emblazoned on them. He holds a sword representing valor. Each gaze intently at the other, ignoring those entering the building, as if the visitors were arriving for reasons other than what each statue represented.

At 10:00 a.m., Staggers called the meeting to order. The first witnesses to speak would be Joe O'Connell and Bobbie Allen of the NTSB. Committee members would have to wait patiently for these men to speak as if they were appetizers before the anticipated main course. Everyone on the

9. "This process is conducted entirely in the open"

committee, and those few covering the hearings in the stands, knew that William McKee, administrator of the FAA, would be served up as the main course.

Staggers wasted no time in explaining the purpose of the meeting. He stated that his committee was not there to investigate the Hendersonville plane crash nor determine the probable cause. He wanted to determine if the current laws applicable in aviation were being followed. He continued, "We are interested in the procedures which are followed and we want to determine if there is anything that the Congress can do to assure ourselves and the people of this country that everything is being done to promote aviation safety."[2] He then introduced Joe O'Connell.

Most of the committee had never met Joe O'Connell and in fact probably did not even know what he looked like. He had been out of government service for many years and had been plucked out of obscurity by President Johnson to lead the NTSB. To some of these young congressmen looking to make a name for themselves, O'Connell looked like a floundering old fish while they, as the sharks, circled and waited for him to flounder so they could go after some fresh meat.

O'Connell began a banal explanation as to how the investigation of all civil aircraft accidents had become the responsibility of the NTSB, transferred from the CAB under the provisions of the DOT act. He went on to state that nothing much had changed in the field of aircraft investigation as a result of the transfer of duties from the CAB. "The practice has always been to organize a team of experts in the various technical areas that might be involved in any such major accident under the leadership of trained investigators representing the NTSB."[3] O'Connell spoke about the exhaustive field investigations that are conducted as well as the scheduling of a public hearing on an accident and the issuance of a final report. "Of course," he assured the committee, "as you know, this process is conducted entirely in the open and as rapidly as facts are identified as incontrovertible and relevant; beginning at the accident site, they are immediately made known to the interested parties to the investigation and are at the same time released to the news media and the public."[4]

He went on to reveal the identities of the interested parties. First and foremost in his mind, speaking as a recently retired CEO of an airline, O'Connell mentioned the concerned airline, Piedmont. He then mentioned the FAA, airline employee associations, airframe and engine manufacturers, and any other interested parties to thus develop a complete record of the facts.[5] Of course, no mention was made of the victims' families or their attorneys, arguably the most interested and concerned of all the parties.

He then updated the committee on the current investigations. He reported that the field investigation and public hearing of the midair collision on March 9, 1967, at Urbana, Ohio, involving the TWA DC-9 and the Baron Beechcraft was complete and the NTSB was now receiving comments and suggestions from interested parties for their final report.[6]

O'Connell revealed that on the day of the Piedmont accident, a public hearing was held on the Delta Airlines training crash at New Orleans, Louisiana, on March 30, 1967. He stated that a final report would be due on that crash also.[7]

O'Connell further informed them that a public hearing was scheduled for the following week on the Lake Central accident near Marseilles, Ohio, on March 5, 1967. He had been the CEO of that airline just a few months before. Member Thayer of the board would preside over that hearing.[8]

As if that was not enough, the field investigation was still under way concerning the Mohawk Airlines accident near Blossburg, Pennsylvania, on June 23, 1967. As soon as the field investigation was complete, O'Connell assured them, a public hearing would be held. His bland infomercial now over, he turned to Bobbie Allen, the director of aviation safety for the NTSB.[9]

Except for the presidential appointees of the NTSB, Bobbie Allen wielded the most power over the Office of Aviation Safety. It seems his interview with LIFE magazine two and a half years before, when he denied that turbulence was even a problem for aircraft, did nothing to call his judgment into question. In fact, even after this shocking and naïve pronouncement, Allen retained his position as the leader of the experts in the Bureau of Air Safety.

Allen knew the spotlight had shone intensely on the NTSB that summer. March had been a devastating month for aviation safety with the Urbana, New Orleans, and Marseilles crashes; everyone was looking for answers. Having come from the CAB, he was part of an organization that promoted commercial aviation as well as investigated its shortcomings. The two goals had obvious incompatibilities, but he knew how many jobs depended on the success of airlines and that it was a duty to protect them as well as the traveling public. That is why the CAB did not vigorously protest the FAA's decision not to ground the 727; that was why planes flew into and out of airports without radar; that was why swift, sleek jets flew in the same traffic lanes as slower, propeller-driven aircraft. The American economy and culture were dependent upon aviation, and the lives of a couple of hundred passengers per year had to be weighed in light of this by those in charge of regulating aviation matters.

9. "This process is conducted entirely in the open" 83

The groundwork had been laid over the last few days with the leaked statements to the press. From the beginning, the focus of the accident was the Cessna. The statements from both the airport director and Harold Roberts from the tower painted a picture of blame, although in Roberts' case it was self-protection. The NTSB refused to contradict those statements. Now it was up to Allen to continue pointing the finger at the Cessna pilot. He could not do it blatantly. Staggers had warned all the witnesses about doing that. Allen would turn the light more brightly on the Cessna while at the same time dimming the light on Piedmont and Boeing. Both companies were in good hands with Bobbie Allen running the show.

The NTSB is the Supreme Court of aviation accidents. Once a decision is made on the probable cause of an accident, it is rarely overturned. The board would look inept if, after spending thousands of hours investigating an accident, as well as hundreds of thousands of taxpayer dollars, it made a determination, only to reverse it later. In order to legitimize its authority, members of the NTSB constantly reiterate their expertise in air accident investigation. Any disagreement with their findings is generally dismissed. This attitude towards those who question their decisions was evident in Mr. Allen's response regarding turbulence a few years prior. Allen knew the audience he now faced, and he used the committee stage to make sure everyone knew that the NTSB knew what they were talking about.

He started with the board's first notification of the accident from a government press officer and how the NTSB immediately called the FAA and the airline to determine the scope of the tragedy. He acknowledged it was abundantly clear that a major tragedy had unfolded, the likes of which the NTSB had not seen in its few months of existence. He then listed those who had departed from Washington National Airport to proceed to the crash site, including Governor Reed and Thomas R. Saunders, the lead investigator for this crash.[10] He excluded from the committee any mention that Saunders was the brother of the vice president of Piedmont Airlines. He obviously did not want to face any questions on that tricky matter.

He went on and spoke about the experts on the board and how the board invited interested parties to help with the investigation. "In every case," he said, "we invited their participation and they responded magnificently by assigning qualified aeronautical experts to assist in the investigation."[11] He left out that most groups in the Hendersonville crash were made up of Piedmont personnel assigned by Piedmont's vice president, who was the brother of the lead investigator. He boasted that "the working procedures of the bureau are so well established, and known throughout the industry, that it is not uncommon for the investigation to be organized and in full operation within a few hours after the accident occurs."[12]

Allen spoke about the areas of interest that the investigative groups would focus on. He listed them as: structures, system, power plants, flight recorder, maintenance records, operations, air traffic control, weather, witnesses, and human factors. He also mentioned that it was the responsibility of Thomas Saunders, as IIC, to formulate these groups and assign the various NTSB personnel to them.[13]

With the preliminaries out of the way, Allen could now turn to talking about the investigation, and at the same time shift any doubt about the cause of the accident away from Piedmont and Boeing and place it right on the Cessna. He revealed that the flight data recorder (FDR) and cockpit voice recorder (CVR) were both pulled from the plane in good condition and flown back to Washington for analysis on Thursday morning, the day after the accident.[14]

Wanting everyone to know that the Piedmont jet functioned normally during flight, he stated that the "flight recorder showed that all parameters, altitude, air speed, heading, vertical acceleration, were functioning in a normal manner. The readout showed that the collision occurred at approximately 2 minutes 37 seconds after liftoff of the Boeing 727 at a mean sea level altitude of approximately 6,132 feet, at an indicated air speed of 228 knots, and indicated magnetic heading of 100 degrees."[15] He added that the FDR further revealed that the 727 was in a stabilized climbing turn prior to the collision.[16] He omitted any mention of the CVR and its recording of the Piedmont crew being distracted by a burning cigarette.

Allen then explained the function of each investigative group and where they were at this stage in the investigation. The operations group is "responsible for developing all facts concerning the history of the flight, before and as the accident occurred."[17] This would include planning, dispatching, medical history, refueling information, etc. He also emphasized the importance of the development of the flight path of both aircraft in coordination with ATC and witnesses in the air and on the ground.[18]

Allen testified that the weather group was responsible for the collection and compilation of all factual meteorological data. Allen explained that the investigators were particularly interested in ascertaining "the sky conditions, cloud coverage, and visibility in the area before and subsequent to the time of the accident."[19] He added that this phase of the investigation was virtually complete. On this point, Allen was wrong. Many of the witnesses to the accident did not complete their statements until at least a week after the accident. Many were still writing what they saw as Allen

9. "This process is conducted entirely in the open"　　85

The shell of Piedmont's Boeing 727 (the *Manhattan Pacemaker*) being pieced back together in Hendersonville during the accident reconstruction stage of the investigation (National Transportation Safety Board).

gave testimony on Capitol Hill. Every indication points to the fact that the weather group had nowhere near completed its phase of the investigation.

He explained that the air traffic control group was responsible for the review of original recordings of the air traffic service, as well as examining all available radar screen recordings and original voice recordings and verifying that written transcripts of voice communications were consistent with the recordings.[20]

Allen further described the witness group as responsible for interrogating all persons who may have seen the accident or had knowledge of the flight or flight conditions.[21] Again, he emphasized the importance of knowing the location of any clouds in the area as well as a determination of the visibility. He commented on the human factors group and its duty to investigate the actions of the crewmembers. This would include any physical conditions and environmental factors that may have influenced the crew.[22]

Allen made specific mention of the structures group, saying that its members were responsible for investigating the airframe and flight controls. He added that it worked in conjunction with the power plant group. That group was responsible for examining the engines, controls, oil systems, etc. Its duty was to determine the existence of any pre-accident malfunction of the power plants or their components. Shifting blame away from Boeing and any problems with the 727, Allen stated, "There is no evidence to date in this investigation that power plant problems were being encountered. It is anticipated that the work of this group will proceed quite rapidly."[23]

In conjunction with both those groups was the maintenance group. Allen said that those investigators held the responsibility to investigate and review all maintenance records of the aircraft. He quickly added, though, that nothing suggested that a maintenance problem was involved in the accident.[24] Allen did not mention the many write-ups about the radar on the Piedmont jet, nor that Piedmont pilots had been calling for that specific aircraft's grounding.

In commenting on the structures, power plants, and maintenance groups, Allen had been methodical in pointing out that there was no problem with the Boeing 727. Less than a week after the accident, with the investigation nowhere near complete, the director of the Bureau of Air Safety testified in front of Congress that there were no problems with Piedmont's Boeing 727. With that testimony, Allen buried any consideration of any problems with the 727 and Piedmont Airlines.

10

"There is nothing in the budget request for radar systems"

With Allen's testimony complete, the committee turned to General William "Bozo" McKee of the FAA. McKee had been here before. As late as the spring, right after the Urbana accident, he had testified about air safety. Those hearings, though, had not been held in public. With the Hendersonville accident and the death of the secretary-designate of the navy and his family, the pressure to have open hearings was more than enough for Congressman Staggers to bear, so he opened the doors of the chambers for everyone to hear.

McKee knew the thrust of today's testimony. He had read the papers over the weekend. He had heard the calls for changing the air traffic control system, of segregating commercial and general aviation aircraft. He knew it was a knee-jerk reaction to a horrible tragedy, but he also knew he would have to answer the questions truthfully and with authority. Not a pilot, concerned more with bureaucracy, McKee decided to bring his deputy administrator, David D. Thomas, with him to answer the more complex aviation questions.

McKee also knew the numbers. He knew the statistics on the accidents and on the money diverted from air safety. Now he sat in front of a congressional committee to defend himself and the president. He was the face of the FAA, and all the decisions went through him. Like everything else in his career, he knew this would be a battle, and he would employ all of his weapons to deflect blame. A strategy of humility, combined with flares of indignation, selective memory loss and excuses would probably get him through the hearing unscathed and with his reputation intact. Most bureaucrats who appeared before Congress probably used the same tactics.

McKee immediately drew a distinction between his organization, the CAB, and the NTSB. Unlike the CAB, which regulates air commerce, McKee explained, the FAA's basic function is the operational aspect of the air system. "We operate the air traffic control system for both civil and military aircraft and provide a system of air navigation facilities designed to promote the safe and efficient movement of aircraft in air navigation."[1] Softening his voice, he expressed dismay at the aircraft accident record compiled during the year. He also acknowledged the ongoing discussion regarding the use of both commercial and general aviation within the same airspace. "Why don't you segregate the two, not only in the airspace but by airport?"[2] he asked, repeating a question that had been asked of him, and the FAA, for a long time.

McKee turned to his expert to answer this pressing question about air traffic control. "At this point, with you permission Mr. Chairman, I would like to ask Mr. Thomas, my deputy administrator, who is generally considered by the aviation community as the leading expert in the country in air traffic control, to discuss this rather complex question. I think it will clear up a lot of questions in the minds of members of the committee."[3]

David Thomas had, for better or worse, been in charge of the nation's air traffic control system for the last decade, even before Congress formed the FAA. He may have been an expert but with mixed results. He commanded the nation's airways when the two planes above the Grand Canyon struck each other, killing 128 people. He also saw the controllers through the midair collision over New York City in December 1960, when a Super Constellation and a DC-8 collided, killing 134 people. Now Urbana in March and last week's Hendersonville crash put the pressure back on Thomas to find out why aircraft were still crashing into each other in the sky.

Thomas started with a basic introduction of the two methods of air traffic control. The first method, he explained, was rules of the road. Pilots use this method when visibility is good, and it is extremely effective, he went on, when airspeeds are moderate and "when the density of the traffic is such that one can deviate from a flight path to avoid another aircraft without, of course, getting into the flight path of another or slowing down traffic. Another method is following instructions from the ground. We use those," he continued, "whenever the weather conditions do not permit the pilot to see and be seen or when the airspeeds, such as those at the high altitudes occupied by the jets, simply do not permit the use of this device. We also use it to insure an orderly flow of traffic in the high-density areas."[4]

According to Thomas's own theory of air traffic control, Addison and

Schulte occupied a "see and be seen" environment. The weather was not adverse. Many witnesses on the ground stated that they saw the aircraft for at least twenty to thirty seconds before the collision and that the skies were clear along their paths. Their altitude at the point of collision was merely 6,000 feet above the ground, which certainly cannot be considered a high-altitude area. With an indicated airspeed of 230 knots at impact, the 727 was not traveling any faster than the capabilities of a piston-driven aircraft of the time. Finally, the Asheville Airport could not be considered a "high-density" area, as no radar coverage encompassed its location.

Now Thomas turned to segregation. "There is a great deal of natural segregation of air traffic in the United States,"[5] he stated. He spoke of altitudinal separation, explaining that jets and turbine-powered airplanes operated most efficiently at the higher altitudes and that slower, piston airplanes stayed at the lower altitudes. Another type of natural segregation is determined by airport. He explained that less than 10 percent of the traffic at JFK Airport is general aviation. However, at Asheville, 64 percent of the traffic is general aviation, and at others, like Rockford, Illinois, 95 percent of the traffic is general aviation.[6] "Our present policy," he said, "is to combine the rules of the road with the necessary ground control as appropriate."[7] Thomas clearly acknowledged that the duty of separation of air traffic is the responsibility of both the controllers in the tower, where available, as well as the pilots flying the aircraft, where applicable. That environment, according to Thomas's own definition, was clearly applicable on July 19, 1967, at Asheville.

He acknowledged that there were many good reasons to seek some segregation between general aviation and air carrier operations. He explained some of the solutions such that the average listener would quickly conclude that the solutions would in no way solve the problem but exacerbate it. One solution, he said, would be to require all aircraft to be controlled by the ground. "At the present time, we are running about 32,000 daily instrument flights in the United States.... If we expanded the present IFR system to require everyone who flies to be under the control of the ground, we would have to expand this number of 32,000 up to some 250,000, which is a manifold increase of our capacity. It would cost several billion dollars and take several years to do so."[8] Another solution, he proposed, would be to stop the airlines operating at any place without radar service. He quickly added that this would stop airline service in about four hundred cities.[9]

In conclusion, Thomas stated that the FAA believed "the air carriers and general aviation are vital to the economy of the United States. The object is not to restrict either, but to provide the facilities and procedures to give maximum safety to both."[10] He acknowledged that there is "no single, dramatic solution to the problem."[11] He admitted that more airports were needed, as well as more reliever airports around the large cities. He also acknowledged a need for more radar and instrument landing systems and said that the FAA was attempting to provide these.[12] He explained that work had been done over the last twenty years to increase the conspicuity of aircraft in the sky with better lights and better paints. He admitted that the FAA was working on enhancing radar returns so that the aircraft structure better reflected radar. He explained the work—and the promise of anti-collision devices for large aircraft against large aircraft—but admitted they were some five years away.[13] He also mentioned the efforts pilots could make to help reduce midair collisions. He stated that the FAA was helping to reduce cockpit workload "and particularly to continue education on scanning and keeping a sharp lookout at all times."[14]

Question time began in earnest, and this time the members of the committee ignored David Thomas. Since they first received the news that another midair collision had occurred in Hendersonville last Wednesday, members had waited impatiently for an opportunity to interrogate William McKee. Now the time had arrived, and the chairman, Harley Staggers, began the questioning, asking the chairman about collision-avoidance systems on aircraft: their necessity, cost, and would they work?[15]

McKee breathed deeply and he acknowledged that intensive work had been going on regarding such a system and that the work was a cooperative effort between the FAA, the aviation industry, and the airlines themselves. He admitted that a workable system was still around five years in the future and that the cost would land somewhere between $30,000 and $40,000 to equip each aircraft. He explained that the airlines were willing to spend the money on such a system but only if it was effective. He pointed out that the system currently under consideration would only be effective between one commercial aircraft and another, not between an airliner and a general aviation aircraft. He conceded that it would not be practical to equip the 100,000 general aircraft registered at this time in the country with such an expensive device. He wanted a system much less expensive that could be installed in all general aviation aircraft. "In all honesty," he admitted, "I must say that this is some time away and we have no breakthrough as of this time in this area."[16] He also added that the system would require the FAA to implement a ground system in conjunction with a collision-avoidance system, which would cost in the neighborhood of $50 million.[17]

10. "Nothing in the budget ... for radar"

If the FAA mandated that all aircraft registered in the country, both general aviation and commercial aviation aircraft, possess this system, what good would it do if the issue was not enforced? The FAA mandated that all commercial aviation aircraft in the country possess weather radar units on their aircraft, but, as evidenced by the unit on Piedmont's Boeing 727, the airlines could flout FAA mandates. Without proper enforcement, any mandate issued by the FAA was just a waste of time. The FAA needed qualified people to enforce its programs, yet William McKee did not want to go that far.

Members of Congress have been accused of acting on certain issues only when it affects them personally. While many of the members knew or had met John T. McNaughton, they had real concerns regarding aviation safety: most were frequent fliers into Washington National Airport, one of the busiest airports in the busiest corridor of air traffic in the United States.

Harley Staggers now turned his attention to radar and congestion. He asked Chairman McKee about his thoughts on the comparison of risk between locations such as Urbana and Hendersonville and Chicago, New York, and Washington, D.C.[18]

McKee acknowledged that there was no terminal radar in the North Carolina accident. Incredibly, he added, "Whether radar would have helped there or not, I don't know."[19] He did admit that one would expect midair collisions in more congested areas, such as the three large cities he named. With the mention of Washington, D.C., other members perked up and started asking questions.

William Friedel, a fourteen-year veteran congressman from Maryland, started his line of questioning. "Would you consider Washington National a very congested area?"[20]

McKee explained that Washington National was an extremely safe airport. He admitted that a lot of flights landed there, but the numbers were the same as they were in 1959 and 1960. "I have no hesitation, and apparently a lot of people in the Congress have no hesitation, or concern, about going into Washington National," McKee responded.[21]

"I think we might disagree on that," Friedel replied.[22]

"I would like to see a lot more people go into Dulles or Friendship," McKee added, interrupting Friedel.[23]

"I would too," Friedel acknowledged.[24] Being from Maryland, he knew the more flights that flew into Baltimore's Friendship Airport, the more money for the state of Maryland. He then complained about the wait time he experienced whenever he tried to use National Airport.[25]

"But, you can't always blame this, Mr. Friedel, on Washington National. Very frequently planes stand on the ramp at Washington National not because they can't get off of Washington National; it is because they can't get in LaGuardia."[26]

McKee confessed that the problem of congestion would only be solved when the government provided better airports, more runways, more ramp space, more taxiways, and more general aviation airports. The airport, he admitted, was part of the air traffic control system. He compared it to radar and communications. "We can't divorce the airport from this problem.... The major bottleneck is the airport."[27]

Several (if not most) of the committee members may have listened in disbelief as McKee blamed Congress for not providing better airports. They knew how much money they had on hand to spend on aviation safety. They were willing to spend that money on aviation safety. They also knew that William McKee had not asked for any money for aviation safety in a long time.

Congressman Paul Rogers of Tampa, Florida, fired the first volley. "General McKee, I understand you said there are about 418 airports that serve commercial aviation that do not have this surveillance radar. Is that about correct?"[28]

McKee agreed that the figure was correct.

"How many airports are served by scheduled commercial aviation that don't even have towers?"[29]

"Without towers?"[30] McKee asked, stalling.

"Two hundred and eighty five, Mr. Rogers,"[31] Dave Thomas chimed in.

Leaning forward, sensing McKee's discomfort, Rogers pressed again. "As I understand it, you have a criteria [sic] where as soon as they have enough numbers of planes coming in and out, this triggers whether they would get a tower or radar.[32]

"That is correct," McKee responded.[33]

"How many airports have qualified under your criteria that do not yet have radar and do not yet have towers? Could you give me those figures?"[34]

Thomas jumped in again. "Under the present criteria that we read for radar, there are some six. On towers there are around fifty.[35]

"When was your criteria last changed?" Rogers asked.[36]

"I believe it was last changed around 1960. It was last reviewed in 1965 and reevaluated. It is looked at about every other year," Thomas said.[37]

Smelling blood, Rogers asked again. "But there has been no change since 1960 in the criteria?"[38]

10. "Nothing in the budget ... for radar"

"No, sir," Thomas admitted. "The tower one has stood the test of time very well and the radar one was changed about six years ago. It was looked at again last year. Correction, 1965."[39]

Rogers knew the answer to the next question. He had researched it over the weekend and had been shocked by what he read. He did not ask these questions for political points. He was not one to posture. He asked because people's lives were at stake. Several of the victims of the Piedmont crash lived in the Tampa area. Now it was time for General McKee and Dave Thomas to answer for that. "What is your budget request for radar systems in the present budget, and for towers?"[40]

Thomas must have known the question would come up in these hearings. He also knew that the answer would shake the very foundation of aviation safety; indeed, if the public knew, the entire industry could suffer.

"There is nothing in the budget request for radar systems or towers in the current budget," Thomas replied.[41]

"Nothing?" Rogers repeated incredulously.[42]

"That is correct," he replied.[43]

Rogers moved quickly to his main point. "As I understand it, your agency is going to get into the SST [Supersonic Transport], which is already at approximately $700 million, if this appropriation is approved, and the House has already approved $142 million," he began.[44] Now that he had McKee and Thomas dangling, it was time for him to gauge their commitment to aviation safety versus American technological prestige. "Do you give greater priority to the SST than to the radar systems and towers that perhaps do something on safety now?"[45]

It had been barely four years since President Kennedy first announced America's intention to build the SST. Now "Bozo" McKee sat in front of the House committee, trying to formulate an articulate, defensible answer on the SST. Johnson had paired McKee and the SST when he announced McKee as head the FAA in July 1965. Since that time, McKee watched the project's support falter to the point that many on the right of the political spectrum did not want it any more than those on the political left. Now Johnson had abandoned the SST and left McKee to support a project that was, in the light of the current aviation problems in the United States, unsupportable.

As if landing a man on the moon was not enough of a national goal, on June 5, 1963, at the Air Force Academy in Colorado Springs, Colorado, Kennedy issued yet another travel challenge to the citizens of the United States: "It is my judgment that this government should immediately commence a new program in partnership with private industry to develop at the earliest practical date the prototype of a commercially successful supersonic transport superior to that being built in any other country of the world."[46]

Kennedy relied on the advice of many in making his decision, but the one man whose backing counted happened to have been the leader of the group assigned to study the feasibility of the SST: Vice President Lyndon Johnson. As vice president, Johnson wholeheartedly endorsed the SST project. His group outlined the costs of the project, the different stages of development, and the requirements of the transport as if the inevitability of the project would not be argued. There was, however, a political element attached to the development of the SST: The British and the French had combined their talents and had started building a Mach 2.2 SST called the Concorde, and there was no way that President Kennedy was going to allow the Europeans to beat the United States in supersonic transport development—especially not after the Russians had beaten the United States into space.

Juan Trippe, the legendary president of Pan American Airways, prompted Kennedy to announce his intentions sooner than he had anticipated. The day before Kennedy issued his challenge at the Air Force Academy, Trippe had announced that he had ordered six supersonic Concordes from the Europeans and he would receive them by 1968.[47] This purchase would showcase European technological advances at the expense of the United States if it decided not to enter the supersonic race. Worse still, the Russians had plans to build one also.

Johnson's committee advanced several points in its support for the SST. The members reasoned that the growth in aviation in the United States had been closely linked to the growth in safety and the speed of the aircraft. They argued that the building of a supersonic transport was important to the welfare of the United States. They considered the transport "the next inevitable advance in commercial aviation."[48]

In keeping with tradition, the American SST would be bigger and faster and fly farther than its European rivals. Johnson's committee recommended that the American SST fly faster than Mach 2.2, preferably at Mach 3 or above. The American plane would also hold 150–175 passengers, as compared to Concorde's 96–104. The range of the Concorde would be approximately 3750 miles, while the American version would be over four thousand miles.[49]

The committee also listed the risks involved in the project. Technological problems, economics, and the sonic boom were the big three. To even enter the supersonic race behind the Europeans and the Russians, the United States knew it had to risk millions of dollars. It was willing to go as high as one billion to beat the Europeans.[50]

Johnson's committee recommended that the project be broken down into three phases: Phase I would be the initial design competition, extending over a period of five months with no cost to the government. Phase II would last approximately one year and would encompass the detail phase submitted by those manufacturers who made the first cut. This would require government funding of approximately $60 million. Phase III would see the development of the airframe through FAA certification by one airframe and one engine manufacturer at a cost ranging between $700–$900 million.[51]

Less than six months after receiving the report, President Kennedy lay dead in his limousine in Dallas, Texas, cut down by an assassin's bullets. Lyndon Johnson inherited the SST program and, having been the primary supporter of its development, felt the need to continue the program. Boeing, Lockheed, and North American Aviation submitted proposals for the airframe design, and General Electric, Pratt & Whitney, and Curtiss-Wright competed for the right to build the engines.

The quietest part of the development of the SST ironically became the loudest. While Boeing, Lockheed, North American, Pratt & Whitney, and General Electric tested their airframe and engine designs, the government ran a series of sonic boom tests on the residents of Oklahoma City, Oklahoma. This was done to find out how amenable citizens across the land would be to faster-than-sound planes flying across their sky belching sonic booms. The citizens of Oklahoma City did not have a choice in the matter. The FAA, aided by the U.S. Air Force, bombarded the sky above Oklahoma City with 1,253 sonic booms for a period of six months, from February to July 1964.[52] The government chose Oklahoma City because of the many ties that community had with aviation. The FAA training center was located in Oklahoma City, and Mike Monroney, senator from Oklahoma, sat in the United States Senate as chairman of the Senate subcommittee on aviation. The government reasoned that most of the residents would not complain because of the city's rich aviation culture. They were wrong.

By springtime of 1964, the eight sonic booms per day had broken 147 windows in the tallest buildings in Oklahoma City. Civic groups sprang up protesting the bombardment and filed lawsuits against the government to stop the tests. A district court judge threw out the lawsuits on the grounds that the tests were a vital national need. The groups then sought a restraining order, which temporarily delayed the testing after it was ruled that the court exceeded its authority. That, however, was quickly overturned, and the tests resumed.[53]

Not surprisingly, it was the noise from the residents, not the sonic booms themselves, that caught the ears of the politicians. According to a study on the Oklahoma City sonic boom tests, "Over 90 percent felt they could accept the eight daily booms. This number dropped to 81 percent during the following eight weeks and to 73 percent during the final seven weeks of the study."[54] Less than five percent of all residents called the FAA office and complained, but in a metropolitan area the size of Oklahoma City, that amounted to approximately twelve to thirteen thousand people.[55]

In 1966, rumors began circulating that more sonic boom tests would be conducted and that the residents Oklahoma City would again be the guinea pig of this aviation experiment. After learning of the thousands of complaints regarding the last test, Monroney knew the backlash on further tests could hurt his reelection chances. He politely but firmly notified the FAA that Oklahoma City would not be subjected to further sonic booms.[56] Monroney's rejection of a second round of testing was the beginning of the erosion of support that the president needed in the Senate for the SST project to continue. Lyndon Johnson's supersonic project was in serious trouble.

In May 1964, the competitors submitted their designs to the SST presidential advisory committee. The committee decided that none of the proposals was acceptable, and they asked Boeing and Lockheed to continue with their designs for the airframe as well as GE and Pratt & Whitney for the engines. North American Aviation and Curtiss-Wright no longer competed to build the transport. In November 1964, Boeing and Lockheed resubmitted their improved designs and waited for a decision.

On July 1, 1965, President Johnson announced that he had received the recommendations of the advisory committee regarding the second phase of the SST project. He remarked that the committee believed there was a high degree of probability that a safe and profitable SST could be built.[57] The president extended the design phase an additional eighteen months, which allowed Boeing and Lockheed, as well as GE and Pratt & Whitney, to continue developing their plans, and he also requested an additional $140 million from Congress to continue the project. At the same time, Johnson introduced McKee as his new FAA administrator. "You know without my repeating it what [your] assignment is. It is to develop a supersonic transport which is first safe for the passenger; second, superior to any

10. "Nothing in the budget ... for radar"

other commercial aircraft; and third, economically profitable to build and operate."[58] Johnson thus publicly saddled McKee with the responsibility for developing the SST. The project was now squarely in McKee's hands, and its success or failure was tied to his success or failure as head of the FAA. Johnson seemed more concerned with an airplane that did not yet exist than with those already in the air.

Congressional support for the project eroded within a year. As the FAA asked for millions more for the project, legislators started to publicly question the reasoning behind the project. The reason most cited was cost. Originally, Kennedy's advisory group estimated that the developmental costs of the SST would be approximately $1 billion. As the fall of 1966 approached, the cost stood at $4.5 billion. "Is this the time to spend federal money on this jet-set frill?"[59] Democratic Senator William Proxmire asked his colleagues. In a recent appropriations vote, Proxmire led the charge against the SST on financial grounds. Senator J. William Fulbright (D-AR) called it an "utterly ridiculous project."[60] Even Senator Robert Kennedy voted to postpone the project, and he was the brother of the president who challenged America to build the plane in the first place.[61] Political watchers in Washington knew the SST was in trouble. Opponents on the political right were hesitant about the government underwriting the cost of a private commercial project; the left about the environmental impact. Both sides of the political aisle had serious doubts regarding the SST, and Johnson knew it.

On New Year's Eve, 1966, General McKee announced that Boeing and GE had won the SST contest. Johnson was at his ranch in Texas during the announcement. When asked about the pending decision a day earlier, he seemed dismissive of the project. He revealed that the administration did not have a definite date for an announcement of the winners, even though he knew full well that General McKee was getting set in Washington to announce the winner within hours. Johnson's political skills were second to none in Washington. He sensed support for the project dwindling, and by the end of 1967 the American SST program was a shambles, unfunded and unloved. It was never built.

"It is not a question of the FAA giving priority to radar or to the SST," McKee replied quickly to Rogers. "The FAA just happens to be the manager of the SST program. The SST program is an administration program,"[62] he added for clarification.

Rogers knew many of the programs within the FAA were administration programs. He also knew money being spent on the SST could be money being spent on radar. "I realize that," Rogers responded. "Radar programs are administration too,"⁶³ he added, making sure McKee knew that he, Rogers, knew the difference.

McKee shifted in his chair. "I am not in a position to say let's cancel the SST program to get more radars,"⁶⁴ he retorted.

Rogers used the SST as bait for the bigger issue he knew lay beneath the surface. He had studied the budget and knew every year what the FAA requested. Now he moved in to reveal McKee's true intentions, and Rogers had a feeling it was not aviation safety. "Let me ask you this question because I don't have too much time. From my understanding of the testimony so far, it seems that this small plane was about twelve miles off. There were cloudy conditions. Wouldn't radar help a pilot in that condition?"⁶⁵ No one questioned from what source he had received his weather report, as the weather group had not even finished its investigation yet.

McKee explained that radar had two purposes. The first was to speed up traffic, which is why it was used mostly in congested airspace. The second was to identify an aircraft's position, and, if it is in the wrong place, the controller can warn the pilot.⁶⁶ "With regard to the Asheville problem, as the chairman pointed out I am in a position now, as the administrator of the FAA, where I cannot appropriately comment on what could or might have happened or what radar would or might have done,"⁶⁷ he answered, avoiding commenting on the Asheville crash.

"I understand that," Rogers added sympathetically. "You don't have the facts. We are not asking you. You don't make a determination of that. I am saying assuming,"⁶⁸ he added hoping McKee would take the bait and reveal the true priorities of the FAA.

McKee bit. "As a general matter with money no object, as I pointed out, obviously there are many areas of responsibility we have; sure, we would like to have radar. But there is a question of priority in the budget, looking at the overall national programs."⁶⁹

"But you have none requested in the budget this year,"⁷⁰ Rogers retorted.

"There is none before Congress,"⁷¹ McKee admitted.

"There are no new ones," David Thomas chimed in, clarifying the confession of his boss. "We have funds for the operation this year for six new radars going in as well as ten towers. But they were all in last year's budget. As far as new radars are concerned, there are none in the budget."⁷²

Several of the members of the committee looked at each other puzzled. They could not understand why the FAA administrator did not ask for the money he needed to safely run his agency without regard to other departments. He was not the director of the national budget. It was not his responsibility to decide overall administration plans regarding all departments. His responsibility lay with the FAA, and they knew he was not requesting adequate funds for his department.

Congressman J.J. Pickle had represented parts of Austin, Texas, since December 1963. Unlike fellow Texan, Lyndon Johnson, Pickle did not care much for secrets. He knew the problems in aviation and knew that a special commission had been created nine months earlier to study airport congestion. He also knew that a special commission had been created to study the problems with the Boeing 727, and he had not heard an answer from either committee.

"Why is it," Pickle began, "that the special commission which has been set up to study airport congestion has not been more aggressive in making some announcement of plans or in holding public meetings, or in making recommendations either to us or to the Congress?"[73]

Pickle continued. "It seems to me that this special commission has been both hesitant, timid, and certainly secretive in making known to the Congress its recommendations.... Why haven't you all made some sort of recommendations?"[74]

McKee quickly responded that he was neither chairman nor co-chairman of the committee. The men who held those positions were Charles Murphy of the CAB and Alan Boyd, DOT secretary. He explained that, as he understood it, Murphy and Boyd had made a recommendation to the president and it was under review. He had no idea when a report would be issued regarding those recommendations.[75]

Congressman Rogers cut in. "Does the FAA have to wait for an outside commission before you can make a recommendation concerning the safety of airports?"[76] he asked McKee.

"No, sir; we don't."[77]

"What recommendations has your agency made?"[78]

"We do all we can in terms of the appropriations we have under the Federal Aid to Airport Act," McKee began, sounding exasperated, as he could not seem to put the issue to rest. "There is an authorization, as you know, of $75 million a year.... Within the resources we have we are taking every step forward we can."[79]

Rogers did not believe him. "I am not very encouraged by that answer. I don't even find a request for more radar and more towers in your budget to the Congress. You are not even requesting them. I don't think that is a very good answer. I thank the gentleman for yielding,"[80] he added, acknowledging Pickle.

Congressman Albert Watson of South Carolina picked up on the secrecy issue from Pickle. Using a golf analogy, he explained that the follow-through is what gives a golfer a good stroke. "Perhaps the answer has been given to the chairman of the committee or to the staff, but as I recall there was an extensive investigation on several 727 crashes. So far as I know, I have never seen any follow-through on the part of those making the investigation."[81]

The CAB had tacked the special report on the Boeing 727 onto the recommendations of the Cincinnati accident report, a report written under the direction of the lead investigator of that case, Thomas Saunders. No one, not Chairman O'Connell, Administrator McKee, or Director Allen, dared reply to Watson. They ignored him.

Congressman Fred Rooney of Pennsylvania eyed McKee suspiciously. He had been in talks with the FAA for a couple of years trying to get radar installed at the Allen-Bethlehem-Easton Airport in his district. He had no luck and knew the airport was about the same size as Asheville's. He did not want to see two planes colliding over Bethlehem. Now he had the opportunity to push McKee on answers that he could not get over the telephone or via letter. "You speak about radar and you see the need for radar," Rooney began. "You wish you could establish radar facilities at every airport in the country. You made that statement, is that correct?"[82]

McKee fidgeted. "I said it would be highly desirable, obviously, if money was not an object,"[83] he replied.

Rooney shook his head. He hated weighing lives versus dollars. "Every reply I have had from your office is always in dollars and cents versus lives. Last year, in fiscal 1967, under the facilities and equipment, how much was requested by FAA?"[84]

"In fiscal 1967?"[85] replied McKee, stalling for time.

"Yes."

"On radar?"

"Well, new facilities and equipment," Rooney responded impatiently. "Doesn't that include RADAR?"[86]

McKee knew exactly where this exchange was headed. "The total for facilities and equipment that was requested of the Congress was $28 million. Our request was $73 million. That was to the Bureau of the Budget."[87]

Rooney did not care about the Bureau of the Budget. He wanted to know how much was requested to Congress. Congress controlled the purse strings, not the Bureau. Rooney saw through the charade quickly. "You requested $28 million in fiscal 1967 and you were granted $28 million."[88]

"That is right,"[89] McKee admitted.

10. "Nothing in the budget ... for radar" 101

Rooney looked at McKee quizzically. "This was $11 million less than you requested in fiscal 1966. So why don't you request more money?"[90]

McKee decided to give Rooney and the committee a brief overview of administration policy. He told them that he was just a part of the whole picture and that he had to take into account what he thought other departments might want. In other words, it was not just the FAA that McKee concerned himself with; it was every executive department in the United States government. "I am only part of an administration. We have to take into account the other problems. We have a problem in Vietnam. We have problems of highway safety. We have problems of marine safety. So there are a lot of things that are highly desirable but, nevertheless, in the priority system we can't justify them."[91]

With this explanation, McKee revealed the priorities of the Johnson administration, with the first and foremost being the Vietnam War. The war was not part of General McKee's problem. It did not fall under his areas of responsibility. Yet he felt that because a non-winnable war raged in Southeast Asia, federal agencies other than the Pentagon needed to limit their spending, even at the cost of innocent lives.

Rooney smirked at McKee. "You are now talking like the director of the Bureau of the Budget. You told me that you needed the money, that we needed the facilities, and yet you don't request the money to install radar installations throughout the country."[92]

Rooney's line of questioning had exposed the FAA administrator as remiss. "General McKee, I would like to ask one other question about the 1966 request of $51 million for new facilities and equipment. You were granted $49 million. How much of that was spent?"[93]

"This was in 1966?"[94] McKee asked, shuffling through some papers.

"Yes,"[95] Rooney responded.

"We asked for $51 million and were granted $49.8 million. Do you want the status of the expenditures?"[96]

Rooney nodded. "You were granted $49 million in 1966,"[97] he said approvingly. "How much was unexpended?"[98]

"I don't have that figure with me,"[99] McKee answered quickly.

Rooney raised his hand as if to let the general know there was no need to worry if he did not have the answer. "It is in the neighborhood of $25 million."[100]

William McKee did not even spend all of the money he was given to develop and improve aviation safety.

11

"I REALLY THINK YOU HAVE BEEN DERELICT IN YOUR DUTY"

By July 1967, the Vietnam War permeated every facet of American society. It ruptured families, caused riots, triggered migrations to Canada, busted budgets, and killed American men. Throwing money into a war that became increasingly unwinnable became the policy of the Johnson administration. The frustrations of ordinary Americans led many to protest against Johnson and Vietnam. The ripple effects of the war became larger, as each week casualty figures increased and American involvement increased. Something had to give. America was being torn apart by Vietnam and was facing its worst internal crisis since the Civil War. Nobody knew that less than Lyndon Johnson and nobody knew it more than John T. McNaughton.

Publicly, McNaughton supported his superiors in their bellicose positions on Vietnam, but privately he knew that the war could not be won. He planned targets for American bombers, while at the same time he worked behind the scenes to stop the whole war.[1] While his public persona supported this surge of money away from many government agencies to be sent to, and lost in, Vietnam, his private efforts attempted to sway his friend and boss, Secretary of Defense Robert McNamara, to withdraw from Vietnam.[2] His failure to do so cost the government billions of dollars and the loss of tens of thousands of soldiers. It also may very well have cost John McNaughton his life.

As John T. McNaughton buckled himself into his seat on Flight 22, the situation in Vietnam was grim. Just two years before in 1965, when the United States sent ground troops in to guard the air force bases whose planes engaged in bombing North Vietnam, McNaughton summed up the

11. "I really think you have been derelict in your duty" 103

John T. McNaughton, second from left, being sworn in as assistant secretary of defense by President Lyndon B. Johnson on July 1, 1964. Also being sworn in as assistant secretaries are, from left, Herbert Miller, Robert W. Morse (third from left), Solis Horwitz, and Daniel Luevano (National Archives Photograph CS14–5-WH64).

war thus far. In a July 1965 memo, he wrote, "The situation is worse than a year ago (when it was worse than a year before that).... A hard VC push is on.... The U.S. air strikes against the North and U.S. combat-troop deployments have erased any South Vietnamese fears that the U.S. will forsake them; but the government is able to provide security to fewer and fewer people in less and less territory, fewer roads and railroads are usable, the economy is deteriorating, and the government in Saigon continues to turn over."[3] McNaughton listed the objectives of America's involvement in Vietnam. He believed that the principal U.S. aim was "to avoid a humiliating U.S. defeat (to our reputation as a guarantor)." To this he assigned the weight of 70 percent. Second, but far less important at only 20 percent, McNaughton believed, was "to keep SVN (and the adjacent) territory from Chinese hands." Coming in a minor third at 10 percent was "to permit the people of SVN to enjoy a better, freer way of life."[4] As in the race for a supersonic transport and the race to the moon, Lyndon Johnson's ministers used the excuse of protecting America's reputation to advance hugely expensive causes with a paltry return on investment for the ordinary American citizen.

While McNaughton had doubts about the United States' ability to win in Vietnam, he still advocated bombing targets in Vietnam. As the

Pentagon Papers point out, in January 1966, McNaughton supported the decision to renew bombing after the president ordered a pause. McNaughton wanted the first strike after a resumption "identified as militarily required interdiction, in order to minimize political criticism. Later strikes could then be escalated to other kinds of targets and to present or higher levels."[5]

Still, for all the talk of placing limits on the war, what with bombing pauses, low-key bombings, and then an escalation, no one placed a priority on the cost of the war. No one at the Pentagon thought to identify the suction sound of American dollars being flushed to protect Vietnam. Nor did anyone seem to care that the war might bankrupt America, or at the very least, force cutbacks in government programs, such as air traffic control, which could put American lives at risk on the home front as well. Lyndon Johnson, though, had to explain, for he was commander-in-chief as well as president. American citizens started to ask why millions of dollars had turned into billions of dollars without a victory to show for it.

At a press conference held at the LBJ ranch at the end of 1966, President Johnson was asked whether America could afford both guns and butter.

> Yes, I think the Nation can afford to continue as we have to fight wars on both fronts.... I think that we must strengthen our people. We must continue our efforts to reduce poverty. We must continue the war against our ancient enemies just as we are continuing it in South Vietnam—until aggression ceases; and until we can provide each child with all the education that he can take, until we can see that our families have a decent income; until we can secure the measures that are necessary to improve our cities, to curb pollution, to reduce poverty. I think this Nation with a gross national product of some $700 to $800 billion can afford what it needs to spend.[6]

Earlier that month, at a press conference Johnson announced another staggering increase in the defense budget. He announced that from December 6, 1966, to June 30, 1967, he would ask for another nine to ten billion dollars for the defense budget. That bumped the budget up to $68-$69 billion for the fiscal year.[7] The war in the air above Vietnam was costing approximately $250 million a month.[8]

As 1967 approached, McNaughton began to waver in his support for Johnson's Vietnam policies, policies that he helped design and implement. He knew the time had come to make a peace in Vietnam and start withdrawing our troops even if it caused embarrassment to the United States. McNaughton addressed the emotional issues the war was causing in the United States. In one of his final memos, he stated:

11. "I really think you have been derelict in your duty" 105

A feeling is widely and strongly held that the "Establishment" is out of its mind. The feeling is that we are trying to impose some U.S. image on distant people we cannot understand (any more than we can the younger generation here at home), and that we are carrying the thing to absurd lengths. Related to this feeling is the increased polarization that is taking place in the United States with the seeds of the worst split in our people in more than a century ...[9]

Someone at the Pentagon finally admitted that the war in Vietnam was not popular and not winnable. McNaughton had often stated that the only way for America to achieve victory in Vietnam was to "lobotomize" the Vietnamese people. In other words, he knew that if elections were held, almost 80 percent of the people in Vietnam would vote for Ho Chi Minh, the leader of North Vietnam. He knew the Saigon government of South Vietnam was corrupt, as did most of the Vietnamese, and that the only way to change their minds was to change their way of thinking. He knew that could never happen; hence, the war was lost.[10] By the spring of 1967, Lyndon Johnson had decided to nominate McNaughton to become secretary of the navy. Why the president considered this move necessary at that time has never been fully explained. Perhaps he wanted McNaughton to have a more prestigious position in the Pentagon. Perhaps, also, Johnson felt that by removing McNaughton as McNamara's assistant, he could distance McNaughton's doubts about Vietnam from Secretary McNamara. Whatever the reason, the Senate confirmed McNaughton to the largely ceremonial position of secretary of the navy in June 1967. After his confirmation, McNaughton took a quick vacation with his wife to western North Carolina to pick up their son Ted from summer camp.

It is doubtful that McNaughton realized, as he waited in the jet's cabin with his wife and son to return to Washington, the extent of the cutbacks to federal programs his advocacy of the Vietnam War had caused. He probably did not realize that there was no radar guarding the Asheville airways and that the FAA had cut back its modernization programs because of the soaring cost of the Vietnam War. He probably did not personally or professionally know "Bozo" McKee and the pressures Vietnam placed on him to make cuts in the FAA budget. There was no way for John T. McNaughton to know that in just a few minutes, he, his son, Theodore, and his wife, Sarah, would become three victims of the cost-cutting measures incurred due to America's involvement in Vietnam.

As the July afternoon wore on, one man sat on the House committee feeling his anger rise at General McKee. Having been in Washington for only two years, he was not used to listening to officials come before Congress and blame others for their own faults. He appreciated Mr. Rooney's and Mr. Rogers' desire to get some real answers as to the aviation problem but was still bothered by their willingness to let McKee get off the hook when the heat got a little too hot. Perhaps it was the Washington game, but Richard Ottinger (D-NY) did not think the lives of the traveling public were game pieces. He remembered not too long ago when two planes collided above his state and killed 134 people. Now, as a congressman dedicated to defending his district and state, he was not going to allow that to happen again.

Nine days before Christmas in 1960, a TWA Super Constellation propeller aircraft and a United Airlines DC-8 jet aircraft collided above New York City. Both had been under the supervision of air traffic control and under radar control. A few minutes before the accident, air traffic controllers had warned the TWA aircraft twice of the approach of the DC-8.[11] One of the two radio receivers in the DC-8 had become inoperative, and the pilots struggled to find their bearings with just one tuner. Flying at almost 500 miles per hour, the pilots did not have enough time to get a bearing on their location, and they overshot their clearance point by several miles and sliced into the TWA aircraft.[12] The Constellation fell on Staten Island and the DC-8 fell on Brooklyn, killing all aboard the aircraft and six people on the ground. This tragedy had happened while the aircraft were under radar control, and now the FAA allowed jets to fly into airports that did not even have radar. Ottinger did not need to be an aeronautical engineer to figure out what would happen under that operating scenario. He had been a captain in the air force and knew a little about flying.

Ottinger had been a supporter of John F. Kennedy in 1960. He became so devoted to Kennedy's ideals that Kennedy hired him as the second Peace Corps employee, the first being Kennedy's brother-in-law Sargent Shriver. After Kennedy's death in 1963, Ottinger ran for Congress and won.[13]

Ottinger looked straight at General McKee, dispensed with the niceties, and said what was on his mind. "I must say, General McKee, that I am not as sanguine as some of my colleagues about the way the agency has exercised its responsibilities in the area of air safety,"[14] he said. "I really think you have been derelict in your duty in terms of actively pursuing the reasonable things that might be done to improve air safety."[15] Very rarely did a congressman use such strong and harsh language towards a visitor to their House. "For one thing, while you are not an independent agency…,"[16] he began until McKee cut him off.

"I am not an independent agency, Mr. Ottinger…. I am a part of the Department of Transportation,"[17] he added, wanting everyone to know that now, due to the recent restructuring of the FAA, others, such as

11. "I really think you have been derelict in your duty"

the Secretary of Transportation Alan Boyd, were also responsible for air safety.

"You are a part of the Department of Transportation, but still you are the only technical agency in the government that is responsible for air safety,"[18] Ottinger countered. Ottinger was not sure if McKee truly understood his responsibilities to air safety. He was about to make sure he did.

"And it seems to me that your responsibility is to define the needs for improvements in air safety, whether or not you can get Bureau of the Budget approval for them,"[19] he added, pointing out the absurdity of McKee's earlier argument. "You are the one agency of the government to which we in Congress and to which the public can look for a definition of what is reasonably needed."[20]

Ottinger continued his lecture in stern tones, glaring directly at General McKee. "If you need more radar, if you need more control towers, if you need more landing lights, if you need more airports, whatever it may be, you should come here and ask for them. It is your duty to do so,"[21] he added, cloaking his argument in military terms. "It is really no answer to say, 'Well, the Bureau of the Budget would not approve an additional request for funds.'"[22]

The congressman continued his criticism, explaining to the general his job functions. "The same thing when it comes to your authority. You say that with respect to separating general aircraft from commercial aircraft, you don't have the authority. It seems to me it is your responsibility to ask for the authority, if you need it, and I think we would consider it very seriously if you did,"[23] he said. "It seems to me that is your definite responsibility to set out for this committee and for the public what is needed to improve air safety. We look to you as the expert for this information. You, as the experts, should define for us what is needed and what could materially improve air safety in the United States."[24]

Ottinger then ticked off several of the areas where he thought the general and the FAA were negligent. He listed lack of continuous ground control in the northeast corridor of the country, lack of radar and ILS capabilities. "I think it is outrageous that the FAA didn't ask for one penny for these purposes this year,"[25] he added. Ottinger continued on about the lack of secondary radar facilities if the primary ones failed and the proliferation of small aircraft around commercial airports and air traffic control problems. He cautioned against the lack of cockpit visibility and the lack of strict rules over air taxis. He reminded McKee that several times during his short time in Congress he had written the FAA regarding these very problems and always received the same answers: "We are working on it," or "We are studying in it."[26]

Clearly reaching his summation, Ottinger wanted to make sure McKee understood his language. "In view of the real crisis that we are confronting, it would require action now. I think you have been derelict in your duties in pursuing these matters more diligently."[27]

By this late in the afternoon, McKee had listened to enough congressmen complain about the job he was doing at the FAA. But with the word "derelict" aimed directly at him and his staff, Ottinger pushed McKee over the edge. With his voice rising, the general responded. "I do not accept your language that I have been derelict,"[28] he said angrily. "Neither I nor my other people in the FAA have been derelict in their duties. I don't know of any 43,000 people in the United States who work around the clock and work harder than the people in FAA.... And,[29]" he continued, "you can give me all the hell you want to personally but I stand up for those people, every speck of the way,"[30] he said, spitting out the last word. "We have got 14,000 controllers working in this control work and working in these centers, and I am proud of them. There are not a finer bunch of men in this country, and if everybody in this country did the job those people are doing, this country would be a hell of a lot better off, and that is my speech, Mr. Ottinger, and I will stand by it,"[31] he added, his voice quaking and heard by all those in the chamber.

Ottinger let McKee's outburst pass without comment. He knew he had touched a nerve in the general. He had done it not for any political points but to try and instill in the FAA chief a responsibility to the flying public's safety. "Well, I'd certainly agree that your controllers are doing an outstanding job under the most trying circumstances. But that misses the point, which concerns the diligence of your agency in pursuing new safety devices and requiring new safety procedures,"[32] Ottinger explained. Then he read a paragraph from an article by the editors of *Electronics* magazine:

> Of all the thousands of government agencies, it would seem almost impossible to rate one as the worst, yet technical men who have studied the air traffic control problems are ready to give that malodorous distinction to the FAA.... The threat of midair collision hanging over the United States when the FAA was formed is still with us, only worse than ever. And the FAA is as far from coping with this threat as it was on the day it was founded.[33]

Ottinger wanted to make sure everyone in the room knew that the critics of the FAA were far and wide, reaching every corner of the United States. "That certainly isn't a very glowing approval,"[34] Ottinger added, not without a touch of sarcasm.

11. "I really think you have been derelict in your duty"

"I would like to answer that too,"[35] McKee replied angrily.

Chairman Staggers knew that without his intervention, this confrontation could get out of control quickly. Staggers did not know the extent to which Ottinger would criticize the agency. As a member of the House for years, Staggers knew it was very rare that such harsh language had been used to criticize the work of a leader of a government agency. Ever aware of the etiquette involved in public hearings, Staggers had to end this showdown quickly. "Just a moment,"[36] Staggers interrupted. "I will allow the gentleman to answer that, then that will be all. You may go ahead and answer it."[37]

McKee nodded his head and acknowledged the chairman. "All I wanted to say is that if I just read the criticisms directed at me and FAA, I would do nothing else and the FAA wouldn't. We would not even have an air traffic control system. I consider that an irresponsible article and the aviation community will back me up 100 percent. I am surprised you even bring it up,"[38] McKee said dejectedly, trying to embarrass Ottinger. Both Staggers and Ottinger ignored McKee's comment, and the responsibility for questions passed on to another congressman.

The rest of the testimony passed less dramatically, with several congressmen trying to soothe the bedraggled FAA administrator. McKee had been around Washington long enough to know that no matter what he said or did from this point on, the press would run with his exchange with Ottinger, and he knew that was an argument he had lost. He had allowed an unknown, two-term congressman from New York to undercut his authority as the primary representative of aviation safety.

12

"A COMPROMISE OF AVIATION SAFETY"

Within a week of the accident, while the aviation bureaucrats argued with the House subcommittee in Washington, NTSB investigators wrapped up the field aspect of their investigation at the crash site. Piece by piece, investigators reconstructed the Boeing 727 at a local gas station in Hendersonville so that they could analyze the fuselage, take measurements and pictures, and perform any other tests that needed to be done on the plane's skeleton.

The investigative teams, disproportionally assigned in Piedmont's favor, met in Washington, D.C., to study the evidence and start writing their group reports, the findings of which would be incorporated into a final report to be issued on the aircraft accident. Most met at the NTSB's headquarters in the Universal building on Connecticut Avenue in Washington, D.C. For several years, the building had housed the Bureau of Air Safety for CAB. It was here that the Bureau of Air Safety had their laboratories, their offices, and all the other tools and equipment they needed to solve plane crashes.

Investigator Ed Patton, a large, portly man who had years of aircraft investigation under his belt, possessed the most impressive array of scientific equipment at the bureau. More an engineer than an investigator, he was the leader of a group that had devised a "reading machine" to decipher information captured on the metallic foil of flight data recorders (FDR).[1] The ability to extrapolate the information from an FDR to read the actions of the subject aircraft before, during, and after impact greatly enhanced the ability to solve the most difficult cases. Patton and his men also used the information from the FDR to coordinate the actions of the crew recorded on the CVR. This capability allowed investigators to reconstruct

12. "A compromise of aviation safety" 111

an accident sequence almost as if they were in the cockpit themselves. The push to equip all commercial aircraft with FDRs and CVRs paid enormous dividends in aircraft accident investigation and in saving lives.

Pacing his office on July 20, 1967, Patton waited impatiently for Piedmont Airlines' assistant director of communications, John Reagan, to arrive.[2] Reagan's flight would not land until late afternoon, and Patton did not expect him until around three or four in the afternoon. It looked like another late night.

The two men had known each other for almost ten years, each serving on the Airline Electronic Engineering Committee.[3] Within the aviation industry, experts in a specific field conducted training sessions for interested aviation personnel. Patton was an expert in FDRs and CVRs. He had been instrumental in pushing the FAA and the CAB to mandate the installation of these recorders in commercial aircraft. Over the years, he met and became friends with many people, the very people whose actions he would often have to investigate: people like John Reagan.

Patton and Reagan spoke on the phone the afternoon of the crash. Patton requested Reagan's presence at NTSB headquarters along with the calibration records from N68650's FDR. He also asked for other records, including those of the navigation equipment, communications equipment, flight directors, and such.[4] Upon arrival at Patton's office on July 20, Reagan handed over all requested documents.

The two men, along with others who formed the FDR investigative team, inspected the FDR system, the tubes, the pressurized chambers, and the magnetic tape. Everyone agreed that the recorder appeared to be in good shape and had been working properly at the time of the collision.[5] Just four days after the meeting, Patton and his group completed their report. It was a short, bland one-and-a-half-page report that stated that the recorder worked properly during the flight and that at two minutes and thirty- seven seconds after takeoff, the collision took place. No analysis was described in the body of the report. Five points were summed up on the last page under Examination and Results. The first read that the "reference line was observed to be consistent." The second read that the "styli operation was normal." The third result was that "all parameters were functioning and recording." The fourth read that the "recording medium was observed to be consistent," and the fifth proclaimed that the "data derived from the readout, as reflected on the attached data graph, appear to be correct...."[6]

Patton only went so far as to say in his report that the FDR worked fine throughout the flight. However, attached to the report in the back was a graph inserted by Mr. Patton. That graph plotted the magnetic heading of the aircraft, and it indicated that just after a minute into the flight, at a height of 4,200 feet, N68650 was put into a left turn.[7] The problem, as everyone on the committee knew at this point, was that the published FAA departure procedures for the Asheville airport mandated that aircraft cannot initiate a turn after takeoff from that airport until reaching a height of 5,000 feet.[8] They would have to figure out a way to explain the deviation, or Piedmont would be in serious trouble.

Tom Saunders appointed John D. Rawson head of the maintenance group. A lead investigator from Miami, he was one of the youngest investigators in the Air Safety Division. He convened his group in Winston-Salem at the office of H.M. Cartwright, assistant vice president of maintenance and engineering for Piedmont Airlines, at 10:00 a.m. the day after the accident.[9] The group was to investigate whether any maintenance issues on the two aircraft led to the tragedy. The group consisted of Ralph Dampier, supervisor of maintenance records with Piedmont Airlines; Lee Gaither, a pilot with Piedmont representing ALPA; and Byron Bodiford of the FAA.[10]

These representatives of the parties to the investigation, intended to help the NTSB examine the accident without prejudice, had the most to lose if they uncovered any discrepancies in the maintenance of the Piedmont aircraft. Two worked for Piedmont Airlines who, if blamed for the accident, could face huge financial losses from lawsuits as well as lost revenue from scared travelers. One of the two was even responsible for maintenance records, and now here he was being asked to sit on a committee investigating his own work. Bodiford worked for the agency that was responsible for monitoring whether Piedmont met and kept the standards allowable for safe passenger operations. Bodiford's reputation, as well as that of the FAA, was on the line.

Clearly the investigative team was made up of those who may have been able to prevent the accident, and now they were being asked to review the evidence against themselves and come up with a verdict. When the group reviewed the maintenance records on N68650, they knew they had a problem. The records indicated that Piedmont management allowed this aircraft to fly almost continuously with radar that did not work. They would have to figure out a way to explain management's decision, or Piedmont would be in serious trouble.

Francis X. Graves, a stout man of a stern countenance, had been with the board since about the same time as Rawson. Tom Saunders appointed him to two committees. His first appointment was as chairman of the Air

12. "A compromise of aviation safety" 113

Traffic Control (ATC) Committee, and the second was to chair the CVR committee.[11] Graves' ATC group consisted of Captain Frank Nicholson, director of flight safety for Piedmont Airlines, and Captain C.R. Malott of Piedmont Airlines. Bennett Bell represented the FAA; Craig Timmerman, the Air Transport Association; and James Daniels, the Air Traffic Control Association. Roys Jones represented the Aircraft Owners and Pilots Association (AOPA), the only person on the committee reflecting the views of general aviation and, by extension, David Addison.[12]

This group faced several challenges. It had to explain why the Piedmont jet turned at 4,200 feet rather than 5,000 feet and whether or not that was permissible. It also had to clear up the confusion of the approach clearance given to Addison by Watkins. It needed to answer whether or not Addison followed his instructions or whether he ignored them and caused the collision. With Addison's views represented by only one person on a committee of seven, the outcome should not have been difficult to predict. Certainly the Piedmont representatives would argue that Addison ignored the clearance, as would the FAA rep and the ATCA rep. The ATA representative's job was to promote commercial aviation. He would side with Piedmont.

The group that would investigate the operational history of the pilots was led by Edwin V. Nelmes, chairman of the operations group. Mr. Nelmes came from his NTSB base in Miami, Florida, to lead this aspect of the investigation. This group would determine whether each pilot followed the operational procedures set out by their respective companies. A main focus of the group would be crew histories, their qualifications, route, and aircraft certifications, as well as their training, especially their training in the Boeing 727. Of the seven members of this group, excluding Mr. Nelmes, over half were from Piedmont Airlines and the remaining three from the FAA, the organization anointed to enforce aviation safety.[13] It did not take them long to realize that Captain Ray Schulte may not have been the best choice for Piedmont to put in charge of a Boeing 727.

It would not do to form a group for the CVR analysis. Too many people with too many interests would take the words spoken on the tape and use them against the pilot, Piedmont and the 727, and Boeing. Saunders had been instrumental in clearing the 727 after the four crashes a couple of years ago with a report stressing pilot accountability: "The board must re-emphasize that the responsibility and authority committed to an airline captain requires the exercise of sound judgment and strict adherence to prescribed practices and procedures."[14] Given Schulte's unsound judgment and breaking of procedure in lighting a cigarette during takeoff, the next words must have haunted Saunders throughout this investigation: "Any deviation can only result in a compromise of aviation safety."[15]

The report also reflected on Zeke Saunders of Piedmont and his management style. "Airline management, too, has a heavy responsibility for devising, developing, and implementing methods and procedures designed to insure that all of their pilot personnel constantly exercise a conservative, prudent, approach to their daily work."[16] In order to clear Piedmont and the 727's reputation, Tom Saunders would need just one man on the CVR committee. He appointed Francis X. Graves as the sole member of the committee that held the most damning evidence against Piedmont.

Tom Saunders had investigated commercial aircraft accidents for a long time. He became an expert on the Boeing 727 during the spate of crashes during 1965 and 1966. He knew that, at the very least, the conversation about the smoldering ashtray recorded on the Piedmont 727's CVR might be a problem. At most, it could derail the entire investigation and place the blame for the accident on Piedmont. It would also resurrect all the bad press surrounding the Boeing 727, the aircraft Saunders himself had deemed "reliable and versatile," with "no major difficulties." He had placed his professional reputation on the line when he gave the 727 a clean bill of health. Would he really be in a position to remain impartial as it became clear that the actions of an undisciplined pilot could put his professional reputation and his brother's company in jeopardy?

13

Reaching a "Breakdown" Point

Alan Boyd had power over all modes of transportation. His dream of a cabinet-level post had come true when President Johnson appointed him the new secretary of transportation. Boyd's experience in aviation, railroads, and highway transportation made him the most viable choice for Johnson. But with the rash of plane crashes over the last several months darkening the traveling skies, he knew he might somehow be held accountable for them. After all, William McKee kept telling Congress that it was Boyd who was co-chair of a blue-ribbon panel appointed by the president to look into airport congestion.

The FAA became a part of the Department of Transportation (DOT) upon the latter's inception on April 1, 1967. The FAA's function within the department was somewhat vague. Congress decided to keep the FAA as an autonomous entity within DOT, but DOT would be responsible for its budget. Just to clarify the matter, Senator Mike Monroney of Oklahoma asked Boyd point-blank during his confirmation hearing about the FAA administrator's role within the DOT: "That man cannot be interfered with politically, in making these final determinations, with the authority to take action?"[1]

"That is correct,"[2] Boyd responded.

Most senators did not think the issue would arise with Boyd. Republicans and Democrats trusted the new secretary. Dwight D. Eisenhower had appointed him to the CAB in 1959, and a few years later President Kennedy appointed him the chairman of the CAB.[3] Boyd knew the current CAB members by their first names, having worked with them for almost ten years. He also knew the newly appointed members of the NTSB. The NTSB investigators, indeed almost everyone at the Bureau of Aviation

Safety, had worked for him at one time or the other in the last decade. He knew the men, knew their families, how they worked, etc. He had been involved with the CAB during many of its most highly publicized aircraft accident investigations. That is why President Johnson appointed him as co-chairman, along with Charles Murphy of the CAB, of the panel to investigate airport congestion.

The blue-ribbon panel, officially called the Airports Task Force, began in 1966. Both William McKee and Charles Murphy advocated to President Johnson the idea of creating a task force to look into the airport problem because the system was reaching a "breakdown" point.[4] The White House accepted the idea but wanted to wait until the Department of Transportation was formed, thus ensuring Alan Boyd's participation in the task force.

Almost immediately, the task force bogged down as it looked for ways to finance the creation of more airports to alleviate congestion in the skies. Some advocated user fees, which the commercial airlines could pay but which the AOPA (Aircraft Owners and Pilots Association), a representative of general aviation, did not support. Some airports could generate enough revenue to pay for improvements themselves, but then state and local boards could get involved and block any bond hearings that might be held for the financing.[5]

The task force also recommended federal loans to meet the financial needs of some airports. However, with some airports already having their service subsidized by the federal government, a change of rules would be required, which could take months or years to make. Others advocated a new loan program headed by Boyd, but funded with user taxes. General aviation was already vehemently against it, but the director of the Bureau of Budget, Charles Schultze, had a different view. "The basic question," he said, "is the extent to which the highly successful aviation industry—primarily serving middle and upper income groups—should pay for the airport costs of the next decade."[6] He further stated that the government needed to review "the premise that the government should underwrite the cost of allowing all private pilots to fly anywhere, and at any time."[7] Of all people, Charles Schultze knew the government was overspending and could ill afford more of it.

Boyd and Murphy secretly turned in their report to the president in early May 1967. It was not released to the public. In March, during a routine hearing on the FAA, McKee was asked by Congressman Rogers what he, as FAA administrator, was going to do about airport congestion. McKee stated that the FAA was studying the problem and making recommendations to the airport task force. However, he admitted that the FAA had not submitted new legislation regarding the funding of the problem.[8]

Rogers was wary of this answer and asked if McKee was waiting for the task force to make the recommendations. McKee replied that he was waiting for the response of the Boyd/Murphy task force.[9]

"Do you know what the time element is on this blue-ribbon committee?"[10] Rogers asked.

"Either June or July of this year,"[11] McKee responded. Within twenty-four hours of his appearance in front of the House committee, TWA Flight 553 collided with a Beechcraft Baron, killing twenty-six people. Aviation safety could not wait until June or July. But it had waited that long, and even though Boyd and Murphy had turned in the report to the president, another crash had happened and this time involving one of the president's advisors. In his heated testimony before Congress in July, McKee had tried to use Boyd's task force to deflect his lack of pushing for more aviation funding.

Now, on top of the blue-ribbon panel, Boyd found he had to contend with NAGE (National Association of Government Employees), a union representing air traffic controllers whose leaders had just publicly called on President Johnson to fire General William McKee and remove him from FAA command.[12]

Holding a press conference in Washington, a union spokesman stated that air safety "has reached crisis proportions."[13] He also stated that "Congress must force-feed the FAA sufficient amounts of money, backbone, and honest concern to act."[14] The spokesman stated that NAGE represented 8,000 of the 14,000 FAA traffic control employees. He explained that the union had begun a campaign last March for more controllers, newer equipment, and lighter workloads in airport towers and traffic control centers. He explained they would send a letter to the president asking him to remove McKee.[15]

Never had an FAA administrator come under so much public fire as McKee. To have a union of eight thousand call for his removal was unprecedented for any executive branch official. Immediately, Boyd came out against the union, calling their demand "completely irresponsible."[16] The Air Traffic Controllers Association (ATCA) sided with Boyd and McKee, stating that they "strongly and vigorously"[17] opposed the idea of removing McKee. ATCA, however, was not a union but a professional trade organization.

Boyd had to have realized that aviation, as it stood at the moment, had reached one of its lowest points in American history. Two midair collisions within a few months had claimed over one hundred lives. Boyd's special task force on aviation sat spinning its wheels in the White House because no one could figure out a way to fund a plan to clear congestion. McKee was blaming the administration and the blue ribbon panel for aviation's problems, and now NAGE was calling on the president to fire McKee. Boyd knew that NAGE's leadership had been called to speak in front of Congress within a few days, which would supply that union with another platform and microphone to broadcast the problems within the aviation community.

It had been easier to fund the space race than it was to fund commercial aviation. Johnson had been a huge proponent of space travel and advocated sparing nothing to make sure the United States led the effort. That, however, had been before Dallas and before the Vietnam War. Lyndon Johnson was a different person now, both in title and in outlook. What Alan Boyd had to do now was go to the White House and warn the president that another midair collision could effectively destroy his presidency.

American airspace is divided into several categories by letter: A, B, C, D, and E. Each category is dependent on the altitude in which an aircraft is flying as well as the proximity to the airport of the aircraft. Category A is the largest area of airspace. It encompasses an area from 18,000 feet to 60,000 feet. Most commercial aircraft fly at this level, and tens of thousands of passengers traverse Category A airspace every day; the lower the altitude, the higher the letters. Categories B, C, D, and E are therefore closer to the ground. There is, though, another area in the sky that is so remote and so secretive that between 1961 and 1967, only a select group of nineteen Americans even had permission from the government to travel through it. That area is called outer space, and the amount of money spent on ensuring safe passage for those nineteen men through that area was far beyond that spent on ordinary Americans who flew in commercial airspace every day.

Lyndon Johnson's legacy in his first few years as president was to fulfill the challenges issued by President Kennedy. LBJ relished the opportunity to compete in a space race with the Soviet Union. He had seen the political implications of doing battle with the Soviets over space domination even as Senate majority leader in the late 1950s. In early 1958, Johnson stated that the free world must control space.[18] Johnson went further, adding, "Control of space means control of the world.... From space, the masters of infinity would have the power to control the earth's weather, to cause drought and flood, to change the tides and raise the levels of the sea, to divert the Gulf Stream and change temperate climates to frigid."[19] Even though he may have engaged in a little fantasy by over-dramatizing the consequences of controlling space, he nonetheless believed it was a goal worth pursuing at any economic cost.

"In essence," Johnson stated, "the Soviet has appraised the control of space as a goal of such consequence that achievement of such control has been made a first aim of national policy.... Our decisions, more often than not, have been made within the framework of the government's annual budget.... Against this view, we now have on record the appraisal of leaders in the field of science, respected men of unquestioned competence, whose valuation of what control of outer space means renders irrelevant the bookkeeping concerns of fiscal officers."[20] Lyndon Johnson wanted to win the space race no matter the cost and no matter the sacrifices needed to meet that cost, and he was not even president yet.

In response to a query from President Kennedy asking for an evaluation of the space program, Vice President Johnson laid out his ideas on America's current efforts in the space program. The vice president admitted that even though America had greater resources than the Soviets for attaining leadership in space, the United States had not utilized those resources. He said that the United States was behind the Soviet Union in the space race and that all nations in the world currently not aligned to either American or Soviet ideals would fall behind the winner of the space race.[21] "This country should be realistic and recognize that other nations, regardless of their appreciation of our idealistic values, will tend to align themselves with the country which they believe will be the world leader.... Dramatic accomplishments in space are being increasingly identified as a major indicator of world leadership."[22]

As vice president, Johnson wanted to win the space race for propaganda purposes rather than for anything the conquest of space might do to make our life on Earth better. "Manned exploration of the moon, for example, is not only an achievement with great propaganda value, but it is essential as an objective whether or not we are first in its accomplishments."[23] He advocated that more resources and effort be put into the space program. He also emphasized the safety of those astronauts involved in the program. "We should move forward with a bold program, while at the same time taking every practical precaution for the safety of the persons actively participating in space flights."[24] The report was given to President Kennedy on April 28, 1961, barely four months after the collision between a TWA Super Constellation and a United DC-8 over New York City, a collision that identified weaknesses in air travel for the millions of passengers flying in American skies every year.

America's first step in manned space exploration began in 1961 with the Mercury space program. According to NASA, the goals of Project Mercury were simple: To orbit a manned spacecraft around the earth, to investigate man's ability to function in space, and to recover both man and spacecraft safely.[25] Except for the loss of astronaut Gus Grissom's spacecraft in the ocean, Project Mercury achieved all of its goals. It did so at a cost of approximately $350 million.[26]

Project Gemini came next. Gemini expanded on the knowledge gained from Mercury, and its goals were precision maneuvering within space as well as extending the stay of man in space, including the "rendezvous of one spacecraft with another while in orbit."[27] The Gemini project also allowed astronauts outside the confines of their spacecraft while in orbit, allowing man to walk in space. The knowledge gained from Gemini brought great strides to the overall goal of the space race: landing a man on the moon. It did not, however, come without a cost. During the span of the Gemini program, 1962–1967, the cost of spacecraft was $790.4 million, launch vehicles came in at $417.4 million, and support came in at $82.3 million dollars. The total cost of Project Gemini came in at a staggering $1.2 billion.[28] One-third of the money from Project Gemini could have bought radar coverage for every airport in the country that needed it.

The task of the Apollo Project was to fly a man to the moon, have him walk on it, and return him safely.[29] The program did not get off to a good start. On January 27, 1967, while undergoing tests for this mission, Astronauts Gus Grissom, Ed White, and Roger Chaffee burned to death in their command module as it sat on the launch pad at Cape Canaveral. Suddenly America's mission to the moon was in jeopardy.

The deaths of the three astronauts led to a swift and massive investigation by both NASA and the government. The United States Senate conducted its own investigation, as NASA, reminiscent of the NTSB's party system, investigated itself. The government virtually shut down the space program. No astronauts flew in space until the investigation was complete and the causes of the fire identified. Never had the deaths of three individuals had such an impact on flight. The investigation concluded that an electric arc in the wiring beneath the capsule had ignited a nearby oxygen panel.[30] Other factors leading to the deaths of the astronauts included not recognizing the test—in a 100 percent pure oxygen environment—as hazardous, and a hatch that required at least ninety seconds to open.[31] The astronauts died as poisonous gases swiftly invaded their bodies.

In 1966, NASA officials estimated the cost of the Apollo program at $22.71 billion. In May 1967 NASA increased that estimate by $472 million.[32]

The increase was due to "the effect of stretching out the Apollo/Saturn V launch schedule."[33] The stretched schedule was due to the deaths of Grissom, White, and Chaffee.

The government's response to Kennedy's call to the space program was unhesitant. The massive amount of funds spent on the handful of men who traveled in space dwarfed all other monies spent on aviation matters. To allocate such a stupendous amount of money on a program that President Johnson admitted was undertaken for propaganda purposes, while cutting back funding in commercial aviation for the millions who flew in the skies every year, put the American traveling public at great risk for the thrill of watching a few men in space suits fly in rockets.

14

"Every major FAA air traffic control facility is short of personnel"

The House scheduled the second day of aviation safety hearings for August 28, 1967, approximately one month after the first session back in July. Those invited to testify included industry insiders such as representatives of ALPA, Aircraft Owners and Pilots Association (AOPA), commercial airline pilots, the Air Transport Association (ATA), and Stanley Lyman and Kenneth Lyons of NAGE, the leaders of the organization that had asked President Johnson to fire "Bozo" McKee.

This time around, only the nine-man subcommittee on Transportation and Aeronautics would question the witnesses. It seems that the full committee had gotten its publicity, especially front-page newpaper coverage of the heated exchange between McKee and Ottinger, so now airline safety hearings would just be relegated to the subcommittee.

One of the first questions to come up was that of a collision-avoidance system between aircraft. The same question had been asked of General McKee back in July. Back then, McKee had explained that the FAA, the airlines, and others inside the industry had for years been optimistic about a collision-avoidance system. He had further stated that the system was in a developmental stage and that it would be three to five years before an effective system could be installed. He added that it would cost $30,000-$40,000 per aircraft and that the airlines "are willing to spend this money if they can get an effective system."[1]

Congressman Samuel Devine of Ohio, representing the state where the March midair crash over Urbana occurred, asked Charles Ruby, chairman of ALPA, about the system. "A lot of people have the mistaken belief that the black nose on the front of the aircraft is radar for everything and not just confined to weather radar...[2] Will it not pick up other approaching aircraft or objects because it is not equipped to do so, that they don't throw enough target back. What is the answer to that?"[3]

"There are two answers to that,"[4] Ruby explained. "One, it is a weather radar. That is its primary purpose."[5] Ruby then went on to explain its secondary purpose, something that, so far, had never been explained. He described the difference between C band radar and X band radar. The C band radar antenna rotates on a 7.5 degree cone. The X band antenna rotates on a 3.5 degree cone depending on which was installed in the aircraft. "Any airplane that is any distance away, the slight tilt of the antennae can completely miss that airplane. So it will not show up as a blip on your radar scope unless perchance you have your antennae tilt set where it will pick up this other airplane, and then the blip is mighty small,"[6] he elaborated. As an afterthought, he added, "You have almost got to know what you are getting a blip from before you know you are getting it from another airplane."[7]

Mr. Devine continued his questioning, apparently ignoring the implications of what Ruby had just stated. "Do you know whether radar research is moving in that direction where they will have a cone type that will detect other aircraft in the area?"[8]

Ruby shook his head. He said that because of the limitations of the cone in the nose of the aircraft, in the fact that you could only see in front of you, an aircraft became vulnerable to collisions from below, above, and to the sides. "The nose type of antennae system simply cannot deal with anything that is behind you or below and behind or above and behind. So there is no point in spending money on researching that for collision-avoidance purposes because you have to cover practically a sphere instead of a limited hemisphere in front of you."[9]

Devine nodded and went on to ask about noise abatement issues and their impact on safety. No one on the committee seemed to realize the import of what Ruby had stated: that the weather radar system was indeed able to detect other aircraft. The weather radar unit was a potential collision-avoidance system, albeit one with limitations. The system's ability to function as a collision-avoidance system was noted by the Boeing Aircraft Company, who told airline personnel interested in its aircraft—in what must have been an extraordinary selling point—that "pilots have reported having seen other aircraft on the scope from time to time."[10]

Prior to purchasing N68650, Piedmont Airlines personnel expressed reservations regarding the radar installed in that aircraft. John Reagan knew before the purchase that N68650 possessed a "C" band radar. He also knew that Piedmont Airlines did not have the equipment to maintain the "C" band radar. He expressed these reservations to his boss, director of communications L.A. Watson, who ignored the concerns.[11] Piedmont leased the aircraft knowing full well that the "C" band radar was unreliable, knowing it did not have the equipment to maintain it, and knowing full well that not only was this a weather radar unit, but that it could be used as an anti-collision device.

The maintenance log on N68650 confirmed Reagan's fears. On numerous occasions, the radar unit did not work. When it did work it was "substandard," "complied with regulations only," and pilots pleaded with management to get it replaced.[12] Piedmont management continued to allow the aircraft to fly. Their decision to fly an aircraft with unreliable radar in an antiquated air system, into airports with no radar facilities, in one of the most dangerous jet aircraft in the world made trips in the *Manhattan Pacemaker* a life-or-death affair.

Unreliable ground radar at the nation's airports exacerbated the problems caused by unreliable airborne radar. Speaking in front of the subcommittee, Stuart Tipton, president of the Air Transport Association, outlined the problems of the current ground radar units.

First, he called the subcommittee's attention to the fact that only 234 of the nation's 526 airports served by scheduled aircraft had control towers.[13] Only 105 of the airports served by airlines had radar service.[14] It was a different figure from the FAA's, but not that much different. In stark contrast to William McKee, who incredulously stated in last month's testimony that he did not know whether radar would have helped at Asheville on July 19, 1967, Tipton testified, "Radar displays in the tower, permitting controllers to observe all traffic operating within their area of jurisdiction, will improve the safety of airport traffic control service."[15]

Tipton also acknowledged that the current radar had many limitations. He testified that the most significant deficiency was its lack of altitude information. "Without altitude information on the radar scope, it is exceedingly difficult for controllers to properly assess the collision potential between aircraft whose blips appear on radar."[16]

He further explained that due to the high ambient light level in control towers during the day, controllers could not see the screens unless shields were placed around them. The shields, though, also interfered with the controller's ability to detect aircraft. He recommended that all radar screens in towers be installed with bright tube displays.[17]

14. "Control facility is short of personnel"

Regarding weather radar in airports, Tipton explained that "thunderstorms and other weather phenomena containing heavy precipitation cause severe clutter to appear on the radar screen to the extent that the screen can become useless."[18] He said, "To offset such problems, circular polarization has been utilized to effectively reduce the display of the radar clutter often associated with severe weather. Where secondary radar is used alone, no weather information is seen on scopes."[19] Lacking proper information, it may be possible for controllers to inadvertently direct aircraft into a severe thunderstorm.[20]

Tipton underscored the main problem facing the national airspace system. He acknowledged the need for additional personnel to install, maintain, and operate the various facilities to handle air traffic properly.

> Virtually every major FAA air traffic control facility is short of personnel. Shortages exist among controllers, maintenance technicians, and installation engineers. The shortages apparently stem from budgetary limitations of recent years. As a consequence, the efficiency of the system has suffered. Burdened controllers are unable to provide additional ATC services which could enhance safety and cope with the growing traffic volume.[21]

Tipton testified in a matter-of-fact manner, checking off the inadequacies of the present system as if reading from a checklist. He neither assigned blame nor pointed fingers. He acknowledged the system was broken and needed fixing. His testimony was broad but not detailed. The individual scheduled to appear after Tipton was more than ready to supply the details. In fact, he had several months ago warned that unless something was done quickly for aviation safety, more crashes would happen.

Certain members of the subcommittee bristled at the mention of Stanley Lyman's name. He was, after all, here to remind the subcommittee of his organization's warnings beginning back in March. The subcommittee, along with the FAA, failed to heed his warnings, and the body count from aviation accidents rose by eighty-two on July 19. Lyman's wrath was not directed at the committee members but at those within the FAA. He had tipped his hand a month before when he called for the removal of William McKee from the FAA. But since they had protected McKee and endorsed his tenure, Lyman felt that the committee should, at the very least, feel a little flak during his testimony.

A year after his inauguration as president, Kennedy signed Executive Order 10988, giving federal employees the right to unionize. Soon thereafter, the National Association of Government Employees was formed by Stanley Lyman's father-in-law, Kenneth Lyons. The seed of NAGE sprouted from the Federal Employees Veterans Association (FEVA). Since veterans organizations could not bargain under Kennedy's executive order, the organization changed its name to NAGE. Within a year, Lyons became president of NAGE and installed Stanley Lyman as NAGE's representative of FAA affairs, raising Lyman to the rank of vice president.[22]

NAGE, however, was an umbrella group representing many different areas of government employees. Air traffic controllers were just a part of the NAGE umbrella. Many air traffic controllers expressed frustration that there was not an organization that dealt solely with their interests.

Another organization representing air traffic controllers was ATCA (Air Traffic Control Association). ATCA was not a union but rather, according to executive director Clifford Burton, an "independent, non-profit organization dedicated to the advancement of science of air traffic control."[23] Many of its members were managers within the FAA, and many controllers felt the ATCA did not adequately represent their interests.[24]

Lyman unequivocally repeated his earlier statements regarding the underlying problem with air safety to the committee. He said the most immediate need was to replace "the managerial people within this agency to correct the serious deterioration which has come about in the Federal Aviation Agency. We feel the only way this can be done is by removing General McKee and his deputy administrator, David Thomas, removed and replaced by articulate gentlemen who have an up-to-date understanding of what is developing within the FAA."[25]

In a written statement prior to his testimony, Lyman repeated the association's warnings over the last several months that all but predicted the Flight 22 tragedy. He reminded them of the extensive brief his group filed with the committee in March, right after the Urbana accident. The brief accused the FAA of mismanagement "coupled with shortages of manpower and equipment that was bringing flight in the U.S. to the point of public peril."[26] He also refreshed them on his rejoinder in March when he publicly called on the committee to hold an in-depth congressional hearing into the entire spectrum of air traffic control procedures within the FAA. In that brief, Lyman went into great detail regarding the number of midair near misses that were occurring throughout the country and the fact that there was a lack of preventative maintenance in electronics.[27] He also warned the committee that "there are insufficient numbers of air traffic controllers in virtually all center and tower facilities to safely handle the rapidly expanding volume of aviation traffic."[28]

Lyman also pointed out several quotes from several magazine articles regarding FAA mismanagement of the nation's airways. *General Aviation News* reported earlier in the year that "at one of the country's busiest and

14. "Control facility is short of personnel"

most important IFR towers, it was necessary for one of the most vital positions to be covered by a nonqualified controller—a man who had never been checked out on covering that position." The article further stated, "This was done with the knowledge of the tower chief and the watch supervisor. It happened because there was no alternative course of action and because there was no one else to do it."[29]

In the December 1965 issue of *Electronics*, the publication commented in its editorial page that "money is only part of the FAA's shortcomings—and maybe the smallest part. Too often, the Agency has tried to freeze technology that was already on the verge of being obsolete…. It has traded in its militancy on air safety for a Casper Milquetoast attitude. FAA men often seem more concerned about rocking the boat than about solving problems affecting travelers on commercial airlines."[30] The article acknowledged that the FAA had tried to improve its control system, but it had failed. Calling into question the president's commitment to bloatedly fund the space race, the writers added that "a technological society in which two space capsules keep a rendezvous somewhere in the immeasurable universe should have no problem getting safe, efficient, comfortable air transportation. But first a major change in attitude and direction is required. The FAA has to do a lot more than count the bodies after an accident."[31]

Lyman then called the committee's attention to a February 1966 FAA booklet entitled "Federal Aviation Agency Cost Reduction." It said the main thrust of the FAA's directives were under an "umbrella" of cost-cutting measures. Lyman added, "We contend, Mr. Chairman, that what we need under such a program are more parachutes instead of umbrellas."[32] Lyman also added, as an aside, that the cost of producing the booklet could have paid for one or two controllers.[33]

None of the booklets or publications, either within the industry or without, had the shock value of the letter that President Johnson wrote to McKee almost a year to the day before the Flight 22 tragedy. Dated July 18, 1966, Johnson's "Dear Bozo" letter expressed his extreme pleasure at the FAA's cost-cutting measures. "I have noted with satisfaction the excellent work which you and your associates at the Federal Aviation Agency have been doing in reducing costs and manpower while absorbing additional workload and improving service to the public,"[34] the letter began. "I have taken particular note of your cost reduction program under which you saved $47 million during the 1966 fiscal year. These savings have been accompanied by a reduction in Agency employment of more than 3,500 employees—eight percent since 1963. The Agency has succeeded in combining economy with a safety program which has helped the commercial air carriers of the United States achieve the best safety record in the world and the best record for any five-year period in the history of American aviation. You have clearly demonstrated that outstanding performance in a critical and complex program can be continually achieved while reducing costs...."[35]

The letter spoke for itself. However, Lyman wryly commented, "The President's confidence that the outstanding performance can go on this way is open to serious challenge."[36] Having now twice challenged the commitment of the president of the United States to air safety, Lyman waited for the committee's response.

Clearly, Lyman stung the subcommittee. After all, it was they who approved the appropriations for the FAA, funds that were not adequate for air safety. They had to defend themselves, and if that meant embarrassing Lyman on a public or private level, they would do so.

The committee's main focus was Lyman's accusation that "electronics people have been handicapped to the extent where they are not allowed to implement a preventative maintenance program to the electronics equipment but rather it is being done as the equipment fails...."[37]

"Would you be a little more specific? You speak of electronic equipment. Give us some examples. You are talking generally. Give us specifics so we can try to follow it,"[38] Friedel demanded.

Devine chimed in. "I agree with that, Mr. Chairman. This is a very serious charge, 'Employees are told not to check equipment.' Now, if you have any information like that, we would like to have names, dates, places, and times, because that is a very serious charge and we would like to know about it."[39]

"We realize this is a serious charge," Lyman replied, "and we say it comes from the electronics technicians themselves who are told 'When you report to X facility to work on radar equipment or whatever piece of electronic....'"[40] Lyman was then cut off.

"If you have names, dates, times, and places we would like to have them,"[41] Devine interjected.

Lyman kept his cool. "Fine. We will be happy to give that to you, sir."[42]

The rest of the testimony followed this basic pattern. Lyman would make accusations, and Friedel and Devine would interrupt asking for names, dates, and places, trying to undermine Lyman's testimony. Then the congressmen resorted to personal innuendoes, casting doubt not only on Lyman's ability, but his mental capacity.

Devine asked him suddenly whether he was a former FAA person.

"Yes,"[43] Lyman quickly responded, cutting off the congressman.

"When did you leave?"[44]

"I officially left on medical separation 60, 65 days ago, but up until that time I was an active air traffic controller."[45]

"Did you leave under any circumstances other than medical?"[46]

"No, strictly medical reasons because of high blood pressure and nerves and I was taking medication that would affect judgment in the case of a radar environment."[47]

Devine had attempted to make Lyman out as some kind of basket case, leaving his job because he could not take the pressure and taking mind-altering pills. The problem with this was that Lyman had been demonstrably right in his warnings to the committee and that the committee and the FAA had failed the flying public.

Lyman next faced searing questions from Congressman Dan Kuykendall (R-TN). Kuykendall tried to play the tough-guy role in his questioning. Referring to Lyman's charge that FAA employees waited until a part broke before fixing it, he demanded "I think the tone of some of the statements made in your remarks would indicate a degree of animosity between your organization and the FAA because at the bottom of page 2 you made a charge, and I am going to ask that this be substantiated—you have charged someone with a criminal act here and I think it should be brought out as to who you mean. I think that you should be able and ready to support that with names. We want them here."[48]

As unfazed by Kuykendall as he had been by Devine and Freidel, Lyman read from an FAA document, dated the same day as the Piedmont crash, which stated, "Operating short of personnel. We may not always have a man available to take care of your problems so be patient and we will take care of the equipment failures as soon as we can."[49]

"You charged the FAA management let a piece of equipment fail,"[50] Kuykendall reiterated.

Lyman explained his point. "If you are in the room and you are working on another piece of equipment and you look across the room and a gear is going to fail and your project for that day is to work on X over here, but Y is failing, you are telling these people, do not work on Y, only work on X. That is correct."[51]

Kuykendall suddenly found himself summarizing Lyman's statement and, in turn, making the most cogent statement by any of the congressmen. "I hope that regardless of what anybody ever tells you that as long as I am flying an airplane around and you look across the room and see a piece of equipment failing that would endanger human life, that no boss in the world would keep you from fixing it."[52]

Lyman realized that Kuykendall had made his point for him, and made sure Kuykendall knew it, too. "Now you are getting to the point that I did not get to because we are trying to summarize and that is exactly what is happening because you have professional and dedicated electronics technicians."[53]

"FAA would let a piece of equipment deliberately fail and endanger human life?"[54] Kuykendall asked.

"In order to satisfy statistics that show that this piece of electronic equipment fails only seldom. Yes, we have that information,"[55] Lyman said triumphantly.

Kuykendall let his anger get the best of him, realizing he had just been manipulated by Lyman, and now fell back on the argument Devine and Freidel had used. "Mr. Lyman, I would like for you specifically to give us the name of one person in the management of FAA that has told directly any person in your organization that they were not to perform critical preventative maintenance that stared them in the face. I would like names. I don't want any more opinions, I don't want any more 'they said,' I want a name and a place."[56]

"I will gladly send to you, Mr. Congressman, the date of the meeting held in New York,"[57] Lyman said; then Kuykendall cut him off.

"I don't want a meeting," he bellowed. "I want a name."[58]

Lyman replied that he would give the congressman the name of the individuals present at the meeting.

Kuykendall had stepped over the line. Lyman had showed the committee nothing but respect and had answered all questions as cordially as he could. Kuykendall must have realized there could be no more excuses coming from the White House or Congress regarding the funding for air safety. Lyman and his organization had warned the committee the previous March about the inadequacies of air safety, and they were correct. Suddenly, Kuykendall acknowledged the underlying issue. "We know that there is a shortage of maintenance personnel,"[59] he acknowledged. "We know that any management of any organization when they have a shortage of personnel has to put in priorities."[60] He acknowledged the failure of the FAA. "We know there is a shortage of personnel and this is one of the purposes of the meeting, to ask for recommendations that the personnel be increased. I happen to be one that is going to join in the recommendation that it be increased, if I have an opportunity to. I will vote on appropriation for it to be increased, but the thing that disturbs me most here about this testimony is the matter of these broad charges without specifics that we in this committee want…,"[61] Kuykendall finished, effectively ending the testimony of Stanley Lyman.

14. "Control facility is short of personnel"

Lyman had done his job. He had finally gotten someone on the subcommittee to admit that there was a shortage of air traffic personnel and that more were needed quickly. Between Ottinger's publicized putdown of McKee, the press conference held by NAGE pressing for the ouster of McKee, and the challenging of President Johnson's commitment to air safety, perhaps Congress would finally heed the calls of NAGE to adequately fund and staff the FAA. If the committee did not follow through, the body count would continue to rise.

15

"A FAMILY AFFAIR"

Technically an NTSB investigation is always open. However, reversals of final reports are rare. Only parties to the original investigation can petition to reopen a case, and then only if new evidence is offered. In other words, once a particular piece of evidence is analyzed and a conclusion reached, that matter is put to rest. It cannot be reversed. Disagreement between parties to an investigation and the NTSB are also rare. In fact, the NTSB reserves the right to relieve any party to an investigation it deems fit. Party personnel realize the futility of questioning the NTSB during an investigation. The threat of being removed from the investigation is too great, and that cuts off any access that particular party would have to information they could use to defend themselves in any civil or criminal suit resulting from the accident.

The uproar over Bobby Allen's choice of Tom Saunders as lead investigator for the Piedmont 22 crash evidently did not bother Allen. He would dismiss any questions about his decision with the same arrogance with which he dismissed the dangers of turbulence. He placed Saunders as lead investigator on this particular accident because of his experience investigating other Boeing 727 crashes. It made no difference that Tom's brother was vice president of Piedmont Airlines. Any party to the investigation that did not like it could leave.

Saunders was an inspired choice for the two parties that had the most to lose in the crash: Piedmont Airlines and Boeing. Boeing loved him because he had helped clear the 727 after its rash of crashes between 1965 and 1966. Piedmont loved him, as most of the airline's senior management knew Tom through his brother Zeke. ALPA would not object because its representatives were Piedmont pilots anyway. The only two parties who might care would be the FAA and the AOPA. But with the facts now public regarding the FAA's delinquency in financing, staffing, and modernizing the nation's airways, the FAA had enough problems of its own and would not be concerned about the lead investigator. And if the AOPA did not like it, Allen could remove it as a party to the investigation. No one playing a part in this investigation would dare question his authority. It must have come as an unpleasant surprise to Allen when columnist Jack Anderson of the "Washington Merry-Go-Round" called and asked if Allen would answer a few questions about the Piedmont crash investigation.

Over five hundred newspapers around the world ran the syndicated column. Read by almost sixty million people, it was the most influential column in the United States.[1] Started by Drew Pearson during Franklin Roosevelt's presidency, it became the bane of Washington power players, including presidents. Several congressmen had been sent to jail after articles alleging wrongdoing appeared with their names in them. Pearson had been the first to report that General George S. Patton had slapped an American soldier during the invasion of Sicily. Douglas MacArthur had tried to sue Pearson over an article, but after Pearson threatened the general with the publication of love letters written by MacArthur to his Eurasian mistress, MacArthur dropped his suit.[2] Only a fool would underestimate Drew Pearson or his partner, Jack Anderson.

Titled "Brother Act," the column about the Saunders brothers appeared in over five hundred papers on Tuesday, August 1, 1967. Anderson called the government's investigation of the Piedmont crash a "family affair."[3] He explained that the government appointed a team of investigators headed by the brother of the vice president of Piedmont Airlines. Anderson wrote, "The government investigators will attempt to determine who was to blame for the collision—the Piedmont plane or a small private plane whose three passengers were killed."[4] Anderson then added why the matter was so important, why any hint of partiality toward one side or the other could have such a profound impact on the investigation. "At stake are possibly millions in insurance. There might be a temptation to pin the responsibility upon the small plane, whose owners are beyond any further harm."[5]

Bobbie Allen had an answer for Anderson's questions: he completely dismissed them. He did not even admit that such a situation would arise or actually happen. He could not admit that there might be someone, somewhere who would want to question the NTSB, or him, about such a thing. In Anderson's column, Allen stated that "the investigation cannot be subverted."[6] He went on to say, "It's a fishbowl operation.... So many officials are involved that no one person, not even the chief investigator, could get away with altering the facts."[7] He admitted that he knew the two men were brothers when the team was selected. "We saw nothing wrong, and we see nothing wrong,"[8] he said defiantly. "No other investigating team was available,"[9] he added, as if daring Anderson to question his authority further.

Allen's responses belied reality. A lot of things could be done with facts other than altering them—Washington, D.C., was ground zero for spin. Present one fact in Congress, and by the end of the day you had so many spins on it that everyone involved forgot what was originally being discussed. More importantly, facts could be ignored. If a fact or—in the instance of the NTSB—a piece of evidence was ignored, the fact would disappear. If a fact was never presented at a public hearing, no one would know the difference.

Anderson's column did not take issue with the investigative team but the leader of the team. While it may have been true that no other investigative team was available, certainly another lead investigator was available. Allen could easily have appointed any of a number of other investigators as lead on this team, but he did not. When Zeke Saunders heard that his brother was the lead investigator of the crash, he excused himself from further participation in the case. That point, however, was moot. If Zeke Saunders excused himself, or even moved to another continent, he would still remain the brother of the lead investigator. The only way to eliminate the issue of impartiality was for Saunders to excuse himself from the case, and this he did not do.

In further questioning from Anderson, Allen revealed that the investigation was broken down into various phases and a group was assigned to investigate each phase of the crash. He admitted that reports from each group would be channeled through Tom Saunders. "The latter will not be able to influence the reports in favor of Piedmont because the hearings will be open."[10]

The hearings would be open, but the reports would also be handled by the NTSB. If a report was not admitted as evidence at the hearing, its contents would not be open. The next time anyone would see the facts, if at all, would be in the final report, which was written by the NTSB months after a public hearing. The only part of an investigation that was open was the hearing, but this was only held for a few hours over a period of several days. Usually, the board member presiding over the hearing would be the board member assigned to the crash from the beginning—in this case, Governor John Reed. But Reed's involvement ended when he left Hendersonville a few days after the crash. Joseph O'Connell then appointed Francis McAdams as the NTSB board member attached to the investigation. McAdams was extremely familiar with the 727, having helped clear it along with Saunders. He was also the man his former boss stated never made a mistake while testifying in front of a congressional hearing.

The implications of the article should have had far-reaching consequences and certainly put the NTSB in a position to relieve Tom Saunders from this investigation. But by the time August 1, 1967, rolled around, the Piedmont crash had been pushed from the headlines, replaced by others just as tragic. Newark, New Jersey, and Detroit, Michigan, had erupted in riots, and the unrest was spreading to other cities. The news in Vietnam continued its downward spiral, punctuated by the horrific fire just days before on the *USS Forrestal* that killed 138 sailors and wounded hundreds more. These were the issues that caught the attention of the public. The fear of American parents whose children were fighting in Vietnam; the images of enraged citizens threatening lives and property: these were more ominous to the American psyche than who was leading the investigation of a plane crash in rural North Carolina.

16

"ERRATIC SPEED CONTROL DURING ENROUTE CLIMB"

The NTSB leaders of the maintenance, ATC, and flight recorder groups must have known that they had problems with the investigation. The maintenance records for the Piedmont 727 could not hide the fact that the radar unit within the aircraft hardly ever worked. The radar system was so bad that Piedmont pilots had begged their management to ground the aircraft. The ATC tapes revealed, contrary to what was being said publicly, that the Cessna pilot, as he flew off Sugar Loaf Mountain, may indeed have radioed his actual flight path to Controller Watkins, indicating that he was going southwest rather than northwest. The flight data recorder showed that the Piedmont jet had turned to the left prior to reaching five thousand feet in direct deviation from FAA departure procedures, while the cockpit voice recorder clearly documented the crew's distraction as they dealt with the smoke from the smoldering ashtray.

NTSB investigator John Rawson led the maintenance group. As the group researched, examined, and discussed the maintenance records of N68650, the history of the aircraft revealed itself. Manufactured in 1963, its first serial number was S/N 18295. Immediately after it rolled off the assembly line, Boeing registered the aircraft as number N7003U. Because it was listed as an experimental aircraft, it again went through a number change on January 23, 1964, this time to N68650. It received its airworthiness certificate on April 27, 1964, under the classification of a Transport Category aircraft. Boeing leased the jet to Iran National Airlines, who returned it to Boeing after the lease was up on October 30, 1966.[1] Since the FAA had no jurisdiction in Iran, no one knew anything about its maintenance during its Middle Eastern sojourn, so the aircraft underwent a four-month inspection by Boeing officials, completed on February 23, 1967.

During the inspection, Boeing examiners concluded that the aircraft had a total flight time of 5555:25 hours. Boeing leased the 727 to Piedmont two days later.[2]

In its report to Tom Saunders, the group stated that Piedmont had supplied all of the maintenance records of the aircraft for review. Rawson and company reviewed all of the records from the first "B" and "C" inspection of the aircraft on March 11, 1967, up until the accident on July 19.[3]

To make the research easier, the group decided to make a list of the different areas that were "deemed pertinent"[4] to the investigation. Of course, at this point, less than twenty-four hours after the accident, no one really knew what had happened to cause the accident, but the group still decided it knew enough to pinpoint the areas of interest. Among the systems the investigators chose to analyze were the air conditioning system, ice and rain protection, landing gear, doors, windows, the oil, and charts. The radar unit was left off the list of systems for the maintenance group to concentrate on.[5]

Pulling straight from the Piedmont Airlines Boeing 727 Maintenance Manual, the group described the various inspections and checks that needed to be performed on the aircraft to keep it flying. The manual gave inspectors a step-by-step list of what to inspect, when to inspect it, and where to inspect it. The manual explicitly stated that each section lists a minimum of the items to be inspected and that maintenance personnel undertake additional items to be inspected if asked and if equipped to do so.

Mechanics performed a service inspection at every stop where the Boeing 727 refueled. They had to check for the quantity of fuel and oil and service each as required. The list dictated that mechanics at the station complete a walk-around of the aircraft, visually checking the cowlings and inspection plates. They also looked at control surfaces, tires and struts, and the tail skid and energy absorber. At overnight stops, the FAA also required a service inspection. Piedmont and the FAA required no forms to be filled out for this inspection.[6]

"The Line Check (Check 'A') must be accomplished within and not to exceed 25 hours flight time or 7 calendar days, whichever occurs first. The Check 'B' is to be accomplished at 100 hour intervals … ¼ of the Check 'C' will be accomplished at each 100 hour inspection period so that at the end of 400 flight hours a complete Check 'C' will have been accomplished."[7]

During these aircraft inspections, the aircraft is broken down into 19 different areas (numbered non-consecutively 11–96) and the mechanics go over each area thoroughly. For example, under Piedmont's rules, Area 11 was the nose, 14 the main passenger cabin, 21 the left wing, 22 the right wing, etc. Different-numbered cards are also issued to the mechanics, with each set of numbers indicating a certain inspection to be performed on a certain part of the aircraft. The cards are also color-coded according to each part of the aircraft. For example, white would be a mechanical issue, blue an instrument issue, pink would be electrical, etc. Each card is only printed once, and a master checklist for the cards has been designed to indicate which cards are to be used for each check. The decks of cards are then brought together and serve as a check-off list to assure completion of the tasks.[8]

An appropriate maintenance schedule is paramount to the safe operation of any aircraft. Certainly, Piedmont's Boeing 727 manual covered all aspects of inspections to be performed, as well as times, dates, and locations. With such an overwhelming number of inspections to perform, with mechanics swarming over the aircraft at almost every stop, it is difficult to understand why Piedmont continued to allow N68650 to fly without working radar, even though on fourteen different occasions in June and July prior to the crash, pilots had written up the radar, several calling it "useless"[9] and another calling for the plane to be "grounded until an operable radar was installed."[10]

Piedmont management knew of the problems with the radar, yet permitted the aircraft to fly. Boeing equipped the *Manhattan Pacemaker* with "C" band radar. According to John Reagan, Piedmont's assistant director of communications and recently assigned to Francis X. Graves' flight recorder group, Piedmont's management knew that the aircraft was equipped with a "C" band radar when it was leased from Boeing, even though Piedmont pilots were not familiar with that type of unit. Reagan had requested that Boeing and Piedmont senior management replace the unit with an "X" band radar prior to receiving the aircraft, but they ignored his request. The reason Reagan requested the change was because after inquiring about "C" band radar with other airlines, particularly United and American, he discovered that it gave more problems than an "X" band radar. The other reasons were more ominous: Piedmont Airlines did not have the equipment to maintain the "C" band radar,[11] and he admitted that the cost to replace a unit was almost five thousand dollars,[12] money that Piedmont did not spend to give their pilots functional radar to transport their passengers safely.

Even with all these problems regarding radar, the NTSB did not mention it in the group report. As a matter of fact, it was completely ignored, and NTSB stated that all inspection forms, "B and "C" checks, campaign

16. *"Erratic speed control during enroute climb"* 139

item worksheets, engineering orders, line checks, performance reports, overnight inspections—in other words, that all maintenance on the aircraft—was performed to standard and was "reviewed, found to be completed, and signed off."[13] The NTSB maintenance group completely ignored the problems with the radar and wrote that N68650 "showed no discrepancies or trends which affected the air-worthiness" of the aircraft.[14] Surely, though, it would be mentioned at the public hearing or in the final report. After all, Bobbie Allen had assured everyone that the facts of the investigation could not be subverted.

Whatever the major problems of the ATC, maintenance, and flight recorder groups were, they paled in comparison to those of the Operations Group led by Edwin Nelmes. As the group delved into the training history of the pilots, it quickly became clear that Piedmont did not heed any of the warnings promulgated by the CAB regarding the Boeing 727 aircraft. Piedmont had no government inspectors qualified for the Boeing 727 aircraft. Check pilots from Piedmont had themselves only been certified for the 727 for a few months, and Piedmont management lacked an independent safety program that could have warned pilots about the dangerous characteristics of the 727.

The first group of pilots chosen by Piedmont to fly the Boeing 727 began training in September 1966.[15] These pilots were the senior officers at Piedmont, including Frank C. Nicholson, Piedmont's director of flight safety, who would, after completing training, conduct line tests on the next group of Piedmont pilots to be trained. The pilots flew to Seattle, where they attended the Boeing Flight Training School to receive instruction in the 727. The training lasted approximately seven weeks and included classroom as well as flight training. Nicholson remembers that "there was a Boeing 727 training manual. We had the Boeing 727 Operations Manual, the Boeing 727 Flight Manual, numerous pictorial pages.... Also portions of the—the pertinent parts of the Maintenance Manual was also in the material which we studied."[16]

There was not, however, any simulator for the pilots to train in. In testimony after the crash, Nicholson said, "I don't think there was any 727 simulator available at that time in Seattle. There was a cockpit and Boeing did have a layout of the cockpit. It was not a simulator, in which you could go and work the switches and do certain things, do certain functions similar to the ones in the aircraft."[17]

After the first group of trainees received their type ratings, a second group and third group were sent to Seattle. The second group received all of its training, both ground school and type rating, in Seattle from Boeing instructors. A third group, though, only received some of its training in Seattle. The rest of it, including type rating, was done within the Piedmont system.[18]

From a safety aspect, Piedmont had not done much to prepare for the arrival of its Boeing 727s, even after being warned by the CAB the prior year that "airline management, too, has a heavy responsibility for devising, developing, and implementing methods and procedures designed to insure that all of their pilot personnel constantly exercise a conservative, prudent approach to their daily work."[19]

In a deposition after the crash, Director of Flight Safety Nicholson fairly admitted that Piedmont had not heeded the CAB's warning. He admitted that Piedmont Airlines had not requested a qualified inspector for the 727 aircraft.[20] When he was asked how many people he had at the time of the accident on his flight safety staff, he replied, "I don't believe I had anyone that you would consider as being on a Flight Safety staff. The only—I believe the only group that would be considered under my supervision at that time was the Operations Control, which handled the scheduling of the pilots."[21]

Nicholson explained that the operations control group was the control office "for scheduling of pilots and in keeping up with their flight times, their trips, trip assignments."[22] When asked by an attorney if the individuals within the operations control group performed safety tasks, Nicholson responded, "No."[23]

The attorney pressed the safety issue with Nicholson. "Did you have any Piedmont pilots who assisted you in your duties as director of Flight Safety?"[24]

"No, sir,"[25] Nicholson replied.

"Prior to the time of this accident, did you formulate as the director of Flight Safety, a Flight Safety program?"[26] the attorney asked.

"No, sir, I didn't have a specific program designated as a Flight Safety program, no,"[27] Nicholson admitted.

"What did you have, sir?"[28] the attorney pressed.

"I don't know that we had anything that was—similar to what you are talking about,"[29] Nicholson confessed.

Incredibly, the director of flight safety for Piedmont admitted that the airline did not have a specific flight safety program, even after being warned by the CAB of the consequences of failing to "exercise sound judgment and strict adherence to prescribed practices and procedures"[30] in the 727 because "any deviation can only result in a compromise of aviation safety."[31]

16. "Erratic speed control during enroute climb"

With the impending arrival of the Boeing 727, Piedmont management should have had plans in place to make sure their pilots had every tool available and that their training did not contain any deficiencies. For most of these pilots, this would not be just a transition to the Boeing 727 but a transition from propeller-driven aircraft to a jet. The difference was truly revolutionary. Speed in a jet was almost doubled from a prop plane, and that decreased the reaction time of pilots. The jets were larger than their propeller counterparts, held more passengers, and required much more concentration to fly.

Piedmont management failed in their decision never to request from the government a qualified inspector of the 727 aircraft to help train their pilots prior to July 19, 1967.[32] Management's inability to formulate an independent flight safety program also reflected their failure to provide their pilots with sufficient support during and after their training period. And with a group of pilots like Ray Schulte, who had never flown a commercial jet aircraft before, coming back to Piedmont's headquarters in Winston-Salem to take over command of the Boeing 727 during passenger and revenue flights, safety support from management should not only have been anticipated, it should have been expected and incorporated.

It should have been apparent to Boeing and Piedmont officials from the very beginning of his training in the Boeing 727 that Ray Schulte had trouble controlling the jet aircraft. Even after he received his type rating, his problems with crew management, instrumentation, and speed control continued. He had taken numerous line tests in the aircraft before he received a passing grade, and on several occasions some of these tests were performed while passengers were aboard. When his final line check was completed after being tested by other Piedmont pilots, some of whom were his friends, he finally received a barely passing grade. While there is no doubt that Raymond Schulte was a dependable and sound pilot during his years with Piedmont in propeller-driven aircraft, his transition into the jet-powered 727 was rife with problems—problems so diverse that had Piedmont followed the CAB's warning regarding the 727, they probably would not have let him fly that aircraft.

Like the other pilots training on the 727, Schulte went through approximately seven weeks of training. The first part consisted of classroom instruction, and then the pilots moved into their aircraft for their type rating at the conclusion of the training.[33] While most of the classroom training occurred in Seattle, the type ratings could occur anywhere. In the case of Schulte, some of his flight training was conducted in Atlanta by Boeing and Piedmont personnel.[34] On April 27, 1967, Schulte's initial flight did not go well. It was so bad, in fact, that the flight check summary written by his instructor stated, "Basic Instrument Procedures (ILS, ETC) not up to level required to pass FAA check, altitude and airspeed control very poor."³⁵

Two weeks later, on May 9, 1967, Captain Schulte received a type rating proficiency check for the Boeing 727 aircraft. Under the watchful eye of FAA personnel, Schulte received a "U" for unsatisfactory in two areas: traffic procedure and holding.³⁶ He had failed his type rating. However, the next day he flew again on another proficiency check and successfully passed in the two areas that just the day before he had failed. Ray Schulte was now eligible to fly the 727.

There are several types of checks that a pilot must undergo to continue flying in his respective aircraft. The first is a type rating check. Once this check is passed, a pilot is qualified to fly that aircraft. Another check is a proficiency check, which is given every six months. According to Piedmont's director of safety, Frank Nicholson, the proficiency check requires that "the maneuvers specified for the proficiency check must be accomplished either in the aircraft or simulator or other means approved."³⁷

Line checks are different. According to Nicholson, "A line check is observing a pilot in the performance of his duties in regular line operations."³⁸ There are no maneuvers to be required of a pilot during a line check, and line checks can be performed with revenue-paying passengers aboard.³⁹

On May 24–25, 1967, Piedmont line check pilot Captain L.W. McNames began checking Ray Schulte on his duties in regular line operations aboard the 727. With revenue-paying passengers aboard the aircraft, Schulte was line-checked on Flights 6, 5, 16, 19, 22, and 33. This area covered most of the mid–Atlantic and southeastern United States. Even though Schulte had received his ratings check on the 727 a few weeks before, he must have regressed in his training. The check pilot observed and noted that Schulte had "erratic speed control during enroute climb."⁴⁰ Enroute climb as defined by the director of flight safety for Piedmont is an area "1,500 feet above the surface."⁴¹ The check pilot also notated that Schulte had "not demonstrated the use of the flight director system."⁴² The flight director system, one of the most important instruments in the cockpit, displays the pitch and bank angles that an aircraft needs to follow a certain route. Also noted by the check pilot was an observation that every captain dreads. The check pilot wrote that Captain Schulte "had not demonstrated positive command up to this point."⁴³ The minimum passing grade on a

16. "Erratic speed control during enroute climb"

line check was 80 percent. Captain Schulte received a grade of 78.[44] He had failed his line check with revenue passengers aboard.

On May 27, 1967, on Flights 4 and 9, again with paying passengers aboard, Ray Schulte was line-checked. This time the line check pilot was Frank Nicholson himself. Nicholson noted that Captain Schulte's "speed control had been improved."[45] However, Nicholson noted that Schulte "was still ragged on his IFR approach procedures."[46] In other words, he had trouble relying on his instruments while approaching the different airports. Nicholson also stated that Schulte's "trip management was better but needed improving."[47] Nicholson also wrote that "crew coordination by Captain R.F. Schulte was better but needed improving."[48] Just like his prior line check, Schulte flunked, scoring a 79, one point below the minimum passing grade.

On May 29, 1967, Nicholson again performed a line check on Schulte. This line check was different. Nicholson noted that Schulte's "climb and departure procedures had improved."[49] He also wrote that Captain Schulte's "IFR approach procedures are getting better."[50] Finally, Ray Schulte passed a Piedmont line check with an 80.

On July 6, 1967, Schulte was given another line check, this time, again, by Captain McNames. While he passed the line check, McNames noted that Schulte "should be more positive in command of crew, particularly with regard to calling for check lists at proper times."[51] Under the Compliance with Procedure section, McNames noted that Schulte performed them "in a satisfactory manner."[52] He also wrote that Schulte's "procedures were generally good. However, failed to set altitude reminder a number of times."[53] Failing to set an altitude reminder means a pilot must rely on his memory regarding certain altitudes in flight. Less than two weeks after this line check, Ray Schulte, along with eighty-one others, died in a midair collision when he made a left turn 800 feet below the required FAA altitude.

In his group report, Chairman Nelmes noted that Schulte had received an unsatisfactory grade on traffic control and holding during his rating flight. Ignoring the failing grades on his preliminary line checks, Nelmes wrote in his report, "He [Schulte] received four preliminary line checks in the Boeing 727 before receiving his initial line check. This was from May 13, 1967, through May 29, 1967. He received a grade of 83 on his initial line check and flew during the period May 13–May 29, 23 hours, 15 minutes and performed 35 landings."[54]

Nowhere in the report did Nelmes write anything about the weaknesses in Schulte's abilities. All the line check reports had been available to, and analyzed by, the operations group, and if there were any questions regarding Schulte's ability, they would have been extremely easy to answer. After all, L.W. McNames, one of the two pilots who had flunked Schulte, was a party member representing Piedmont on the operations group. But maybe, just maybe, the party system would work. Perhaps Nelmes would ask McNames about Schulte's training, McNames would answer, and Nelmes would put it in his report. However, if that happened, the NTSB would have to bring to light Piedmont's pilot training procedures and acknowledge in its report that Ray Schulte was perhaps not the best pilot for an airline to put in command of a Boeing 727.

17

"That's just the cigarette that's on fire"

Francis X. Graves listened to dead people for a living. Through cockpit voice recordings, he heard the last minutes of people's lives, the chatting and conversation among people unaware that death is just moments away. Sometimes the recordings were clear; others were not so clear. But the recording from the *Manhattan Pacemaker* made it plain that, just before they died, the crew talked about a smoldering cigarette in the cockpit. David Addison and crew, though, were not as articulate as the Piedmont pilots. It would take weeks of playing back tapes, asking for interpretations, and even requesting help from other government agencies before Addison's voice was heard on tape specifically notifying James Watkins of the exact heading he was flying coming off Sugar Loaf Mountain.

As N3121S approached the Asheville VOR on top of Sugar Loaf Mountain, the air traffic nerve center in Atlanta had the plane under control. At 11:50:45 a.m., personnel in Atlanta cleared Addison to the Asheville VOR and told him to expect an ILS approach at Asheville.[1] At this point Addison, with over ten thousand hours of flight time experience, obviously pulled out the ILS approach plate for directions on how to conduct an ILS landing in Asheville. A man of his experience must have done this a thousand times at hundreds of airports. Just under four minutes later, Addison probably plainly heard the Asheville tower clear Piedmont Flight 1022, Paul Snell's Martin 404, for an ILS landing into Asheville, as both had their radio tuned to the approach frequency.[2] It was obvious to David Addison that ILS landings were in effect in Asheville.

At 11:56:27, James Watkins issued his confusing clearance to Addison. "Three one two one Sugar, cleared over the VOR to Broad River. Correction; make that the Asheville radio beacon. Over the VOR to the

Asheville radio beacon. Maintain seven thousand, report passing the VOR."[3]

At no point did Watkins tell Addison what approach to use at Asheville. FAA regulations state it is mandatory that a controller will transmit upon first radio contact or as soon as possible thereafter the type of approach clearance or type of approach to be expected if there are two or more approaches published. At Asheville, there are four. The Asheville radio beacon was not depicted on the ILS approach plate to Asheville.[4] Had Watkins advised Addison what approach to use on first, second, or even the third contact, and by each of those times Watkins knew he wanted the Cessna coming in on an ADF-2 approach, the collision would have been averted.

Complicating things further, Watkins, in violation of another FAA rule, did not advise Addison what heading he wanted him to fly off the Asheville VOR. From the Asheville VOR, the Asheville radio beacon is located to the northwest. The Broad River radio beacon is located southwest of the Asheville VOR. FAA regulations require that a clearance to a non-directional beacon include the course to or bearing from the facility.[5] The Asheville radio beacon is a non-directional facility. Had Watkins followed this procedure and given Addison his course off the Asheville VOR, the accident would not have happened.

At 11:58:21, James Watkins picked up the interphone line to Atlanta and told the controllers there of his traffic situation. Nine words into his conversation, David Addison radioed his heading. "Two one Sierra, just passed over the VOR. We're headed for the Oh Good Shit Via Two Three Eight for uh Asheville now."[6] After he finished his conversation with Atlanta, Watkins replied to Addison: "Two one Sugar. By the VOR, descend and maintain six thousand."[7] James Watkins acknowledged during the public hearing that he did not hear David Addison's transmission. Yet, inexplicably, he cleared Addison into airspace soon to be occupied by a climbing 727. Clearly, James Watkins did not perform his duties that day the way he was trained to do. His failure to request that Addison repeat his transmission led directly to the collision.

NTSB investigator Francis X. Graves did not understand Addison's transmission, and he had listened intently to the tapes for days after the accident. Even though the NTSB had very little filtering equipment available, Graves spent a period of almost three days listening to the tape. "So primarily," he stated when asked of his account, "it was just work with the tape, listening to it, headphones, varying volume levels, varied speeds, and repeated listening to it."[8] After this three-day effort, Graves still did not fully understand what Addison had said.

Graves asked Bobbie Allen to contact the Federal Bureau of Investigation (FBI) and ask them for help. Allen sent his letter on August 30, and Robert Rudick of the NTSB delivered a copy of the tape to the FBI on September 12. On September 27, the FBI responded to Allen with a memo stating they had applied various filtering techniques and used a sound spectral analysis in an attempt to understand the words. The memo bluntly stated that "the examination failed to reveal any recognizable intelligence, and insufficient detail was present to permit identification of the questioned sounds with the pilot's voice."[9] Unfortunately, the FBI failed to recover Addison's words.

The FAA, though, also wanted to know what Addison said. Bernie Curtis, a manager in the Air Traffic Service of the FAA, contacted the National Bureau of Standards, with Graves' knowledge, and asked personnel there to analyze the tape. Edwin Burnett, a physicist with the sound section of the Institute of Basic Standards, a part of the National Bureau of Standards, analyzed the tape. He recorded the garbled section of the transmission and made a loop of it so it could be listened to repeatedly without any inconvenient stops and starts. Burnett described the process. "A small improvement was made by slowing down the tape slightly. Some further improvement was made of selective filtering of various frequency bands...."[10] Applying these techniques, it did not take him long to figure out what had Addison said.

On September 28, Francis Graves, along with members of the FAA, arrived at Mr. Burnett's laboratory on Connecticut Avenue for an explanation of how he obtained the words he reported were on the tape. For thirty minutes, Mr. Burnett described the techniques he used for his analysis. He told Graves that he was not the only person to come up with these words, that several others in the lab agreed with his findings. Burnett told him that the first three sounds were understood to be "O good shit." After repeated listening, Burnett and several others in the lab heard Addison say, "Via two three (or two) eight. Via seems to be preferred to right and three to two.... These phrases could have been other words with the same cadence and similar sounds. Those given seem the most probable."[11] If a pilot flies off the 238 dial of the Asheville VOR, he is clearly on his way to the Broad River radio beacon. Addison had told Watkins that he was heading southwest towards Broad River, not northwest towards the Asheville radio beacon. Watkins never heard the transmission because he was on the phone with Atlanta Center. His failure to ask Addison to repeat his heading off the Asheville VOR led directly to the Cessna flying right into the path of the *Manhattan Pacemaker*.

On October 5, Edwin Burnett wrote his findings in a memo and had it delivered to Bernie Curtis. The next day, on the sixth, Francis Graves amended his ATC factual report, fully aware that Edwin Burnett of the National Bureau of Standards had deciphered the tape. Graves simply ignored Burnett's analysis. Graves wrote in his addendum, "Efforts to gain intelligence from the garbled transmissions on the tape recording produced no conclusive results, although three or four sounds were thought to be intelligible words."[12] He added that he had sent a copy of the tape to the FBI for analysis and that the FBI examination "failed to reveal any recognizable intelligence insofar as the subject voice transmissions are concerned."[13] Francis Graves ignored Edwin Burnett's analysis of the tape and completely refused to acknowledge that the National Bureau of Standards had even analyzed it. Surely he would change his mind and mention it at the public hearing and include it in the final report.

There could be no ambiguity about what the crew in the cockpit of Flight 22 spoke of just seconds before they died. The CVR had recorded their final conversation about smoke in the cockpit of their plane. Investigators transcribed their words, revealing an incredible lapse of professional judgment by the captain, who allowed a lit cigarette to catch fire in his ashtray and distract all of them from performing their duties. Moments later, the Cessna 310 plane struck their aircraft. According to the NTSB, the Cessna had been visible for almost thirty seconds prior to the collision.

A Fairchild CVR, model A-100 serial number 485, had been extricated from the wreckage approximately six hours after the accident and shipped to NTSB headquarters in Washington. Investigation revealed that the tape held four recording tracks that belonged to the cockpit area microphone circuit, the captain's radio circuit, the copilot's radio circuit, and the engineer's radio circuit. Analysis revealed that the recorder was operating normally when the collision occurred.[14] No consensus was taken from the CVR group on these matters. None was necessary. No group had been assigned to this critical part of the investigation. Francis X. Graves had the distinction of being the sole member of the CVR group.

For years, especially with the addition of jets into the same airspace as propeller and general aircraft, the flying public's risk of midair collisions increased. Pilots had to be alert and on the lookout for other aircraft at all times. The concept was called "See and be seen." Throughout the 1960s, the NTSB expected that looking out for other aircraft will remain the "basic means of collision avoidance."[15] The board admitted that the air traffic control system could not provide positive separation of all aircraft at all times. The board urged that "all users of the airspace make every effort to achieve the maximum benefit from visual detection. No less than constant vigilance on the part of both pilots and controllers is required."[16]

Piedmont Airlines knew about the growing danger of midair collisions. In an Operations letter dated February 28, 1961, management alerted pilots to stay at the controls of their aircraft when it was taking off, landing, or en route because "increasing air traffic and the resultant increased possibility of mid-air collisions require constant vigilance and attention to duties by all crew members."[17]

A year later, in an Operations letter dated April 25, 1962, Piedmont addressed the issue of cockpit access: "We all know that cockpit vigilance is now of paramount importance because of the congested traffic conditions and the speed of our modern aircraft on our airways. The above conditions require constant surveillance on the part of flight crew members...."[18] Signed by W.O. Tadlock, Director of Flight Operations for Piedmont Airlines, this letter was circulated before Piedmont had received any jet aircraft. Even with turbo-jets and propeller aircraft, Piedmont urged its crews to maintain a proper lookout for other aircraft, even using words like "constant vigilance" and "paramount importance."

When Piedmont decided to enter the jet age, it chose the most dangerous jet being offered for purchase, the Boeing 727. Three of them had crashed within a period of a few months in the latter part of 1965. In January 1966, after the fourth crash, the FAA summoned all operators of the 727 to Washington for an emergency meeting of the aircraft. The CAB also performed a special study on the aircraft's operational capabilities and capacities. It tagged the results of this study onto the final report of the American Airlines Flight 383 crash on November 8, 1965, outside the Cincinnati airport. Tom Saunders led the investigation of that crash, as well as being a prominent contributor to the 727 study. He staked his professional reputation on the Boeing 727 being a safe aircraft.

The results of the study reported that the 727 "*does have* highly responsive and versatile flight characteristics and that these favorable characteristics may be misleading to the pilot, or are presenting the impression that greater liberties may be taken with the aircraft in normal operating situations...."[19] In other words, if you flew this aircraft, you better pay complete attention to your duties.

The board also reemphasized the responsibility and authority of an airline captain. When flying any aircraft, but in particular the 727, the pilot must "exercise sound judgment and strict adherence to prescribed practices and procedures. Any deviation can only result in a compromise of aviation safety."[20] The CAB wanted to go a step farther with that aircraft, though.

It wanted airline management involved in and observant of pilots' actions. "Airline management, too, has a heavy responsibility for devising, developing, and implementing methods and procedures designed to insure that all of their pilot personnel constantly exercise a conservative, prudent approach to their daily work."[21] The CAB would now hold management responsible for pilots' actions.

The CVR tape had recorded the crew from the time they were on the ground in AVL, through takeoff, to the time it took to strike the Cessna. At thirty-seven seconds past noon, flight engineer Lawrence Wilson smelled something odd. "Somebody got an ashtray on fire?"[22] he asked. At that point, all cockpit discipline broke down. There can be no question that with the word "fire," all three men looked for the source.

Captain Schulte looked down at his ashtray located on the side of the cockpit by his left arm. "I do, I think."[23]

"Okay,"[24] Wilson replied two seconds later.

"You know it couldn't be me,"[25] Conrad chimed in, continuing the banter.

"Ashtray isn't on fire. That's just the cigarette that's on fire,"[26] Schulte determined. Seconds later, Schulte extinguished the smoldering cigarette.

"I'm sorry. I fucked up again, didn't I?"[27] Wilson replied. "Just for that, I'll burn your damn steak,"[28] Wilson said. It took nine more seconds before the next words were uttered regarding their flight duties. For almost thirty seconds, the crew bantered about the fire, not paying any attention to the sky outside.

Captain Schulte's smoking while engaged in the busiest time of a flight was careless at best. Allowing the cigarette to smolder in the ashtray made the situation worse.

According to Tadlock, director of flight operations, Piedmont's written policy in its Operations Manual states the crew "shall not smoke in the cockpit if the cockpit door is open and where it would be visible to passengers."[29] Captain Tadlock, however, divulged more. He also admitted that the general practice of Piedmont crews is that "they don't smoke when the passengers are not allowed to smoke."[30] Tadlock finally confessed that from the time Captain Schulte lit his cigarette, to the time he let it turn into a distraction from his duties, to the time of the collision, the No Smoking sign was on for the passengers.[31]

Five seconds after Wilson smelled smoke in the cockpit, the Cessna 310 could have been seen by the crew had they have been looking directly at it. The Cessna could be detected at 20 degrees at a range of about 7100 feet had they been looking out the cockpit window.[32]

17. "That's just the cigarette that's on fire"

The cockpit vigilance required by the CAB and the NTSB evaporated in the cockpit of Piedmont Flight 22 when Captain Schulte lit his cigarette, setting in motion a chain of events that included lack of proper scanning techniques, a failure to maintain a safe lookout, violation of company procedure, and pure carelessness. Now it would be up to the NTSB CVR member, the lone member Francis X. Graves, and Thomas Saunders as the IIC to decide whether this blatant breakdown of cockpit vigilance and discipline should be cited as a cause of the accident or be ignored. Surely it would be included at the public hearing and in the final report.

18

"We'll turn off, go direct to the VOR"

For airline executives, the end of the 1950s was a time of decision. Airlines had to decide whether to convert their fleets to jets or stay with the piston-driven aircraft. It was as important a decision as most managers would make in their careers. As the 1960s approached, airport owners also faced an important decision. Should they convert their airports to accommodate the jets or stay with the pistons? That decision could affect the economy of the entire area the airport served. Without a useable airport, it would be nearly impossible to attract businesses and industries to invest in the area.

By 1957, the leaders of the Asheville and Hendersonville communities decided they needed a new airport. The decision to build a new airport was based solely on the proximity of the current A&H airport to the mountains. A&H sat virtually astride and inside a series of mountain peaks that made aircraft operations hazardous even on the clearest of days. There was also no room to expand the runway to make it longer to accommodate the approaching jets. "The A&H airport was limited due to the proximity of mountain ranges surrounding the campus, so a new location and airport was planned by leaders in Asheville. In order to grow the airport to accommodate larger planes used in commercial air travel, a bond issue authorizing the City of Asheville to expend up to $1.2 million was passed."[1]

On January 15, 1961, the FAA approved commercial aircraft to operate to and from the Asheville Regional Airport, and on June 7, 1961, the terminal building opened.[2]

As with any new airport, approach and departure procedures had to be written by the FAA. That agency writes regulations regarding virtually every aspect of flight. The published regulations are readily available to pilots in many ways. They are posted on approach plates, listed at airports, located in any library, and they can be purchased. Federal Regulation Section 91.87 covers operation of aircraft at airports with operating control towers. Under departures, the FAA promulgated, "No person may operate an aircraft taking off from an airport with an operating control tower except in compliance with the following: Each pilot shall comply with any departure procedure established for that airport by the FAA."[3]

In 1961, Earl Cato, chief of air traffic control in Asheville, requested a change in IFR departure procedures to the south of the airport. As the airport opened, it was procedure to climb out to the Broad River Intersection at five thousand feet or higher before proceeding on course. Cato wanted the change to indicate that as long as an aircraft reached five thousand, whether before the Broad River Intersection or not, it can then be required to proceed on course.[4]

Cato wanted this changed to allow air traffic to get into and out of Asheville quicker than the rate it was currently operating under. He wrote, "The above change would expedite the flow of traffic in the Asheville Terminal Area by permitting Air Traffic Control to make more efficient use of diverging routes and to permit the outbound aircraft on south takeoffs to expedite clearing the one approach area that is available to inbound aircraft."[5]

The departure procedures had nothing to do with terrain avoidance; after all, that issue had been put to rest when civic leaders moved the airport miles away from the mountains. As a matter of fact, the only mention of terrain around the airport was a caution placed on the FAA procedure. It read: "CAUTION: Terrain rises rapidly 2.0 miles west of the airport."[6] Printed on the ADF-1, ADF-2, and ILS approach charts for the Asheville airport is the following procedure for aircraft taking off to the south. "Takeoffs to the south will climb on course 161° over the OM (Outer Marker) and continue on course 161° to Broad River Radio Beacon. Upon reaching 5,000 (feet) or higher as directed by ATC, continue climb on course."[7] While Cato's exact wording may not have been implemented, certainly the height requirement was the same. A pilot cannot turn his aircraft off course 161° (which is the runway heading) until he reaches a height of 5,000 feet.

As Piedmont Flight 22 waited impatiently at the end of runway 16 to begin its takeoff roll, Captain Raymond Schulte was obligated by FAA regulation to adhere to the above departure restriction. He could turn on course for Roanoke, Virginia, only after he reached five thousand feet. Moments before, while Flight 22 was taxiing to the runway, the tower issued it an IFR clearance in accordance with the one stored in the aircraft's computer system.

"Piedmont 22 is cleared to Roanoke Airport via direct Valdese J fifty-three flight plan route. Maintain flight level two one zero." This message reached the crew from the tower at 12:54:12.[8] Ten seconds later, the crew repeated the clearance.

Less than two minutes later, the crew of the Boeing 727 advised the tower they were ready to go, but the tower advised them to hold their position at the end of the runway. The crew's impatience began to show, as they were still trying to make up the time they had lost in Atlanta earlier that morning.

At this point, Paul Snell entered the conversation as the captain of Flight 1022 approaching from the southwest and descending from seven thousand feet to six thousand feet. For the first time, the tower realized there might be a potential problem between the two Piedmont flights.

Incredibly, with no involvement from the tower, Schulte and Snell worked out their own separation problems on their company radios. "Cancel the damn thing, Paul,"[9] Schulte said, urging Snell to cancel his instrument approach and proceed visually. Eight seconds later, Snell did as he was told. "OK; 1022 would like a contact approach."[10]

Rather than telling Snell what to do, the tower asked him what he would like to do. Snell responded that he would come up out of the way of the jet.

Hearing Snell's answer on the radio, Schulte chimed in. "We'll turn off, go direct to the VOR."[11] Schulte did not realize that just over a minute before, approach control had given Addison in his Cessna 310 the convoluted clearance off the same VOR. The tower also did not bother to tell Schulte of the Cessna's approach. In fact, a few seconds after this, at 11:58:01, the tower cleared Flight 22 for takeoff with the caution to maintain runway heading until reaching five thousand feet.

The tower restriction of five thousand feet, according to controller Jesse Welch, was placed on Flight 22 because of the Cessna.[12] One minute and fifteen seconds after Addison reported over the VOR, the tower released Piedmont from its restriction. "Piedmont 22, climb unrestricted to the VOR; report passing the VOR."[13] This restriction, however, only dealt with traffic separation, as 3121S was approaching from the east. The pilot is still responsible to follow the FAA published procedures, which indicate that an aircraft may not turn before five thousand feet.

At 4,200 feet, when the tower cleared the Boeing 727 to climb unrestricted to the VOR, the air traffic controllers did not have the technology

18. "We'll turn off, go direct to the VOR" 155

or capability to know the altitude of the jet. That responsibility fell on the pilot, and it left Ray Schulte with a decision: should he continue on course to 5,000 feet, a course that would last only a few seconds more, or should he take the shortcut to the VOR and turn before he reached 5,000 feet? When Schulte decided to turn, he did not exercise the sound judgment and strict adherence to prescribed practices and procedures that the Safety Board stated must be followed when flying a 727. He decided to turn prior to the FAA's published departure procedures in the sky around an airport that did not have any type of radar coverage, in an aircraft with unreliable radar, in an area where, if just one pilot is confused about any clearances received from ATC, or another does not follow the rules, disaster could appear. Surely, the decision to turn prior to five thousand feet would be cited as a cause of the accident.

19

"I DID NOT PARTICIPATE IN THE NTSB'S INVESTIGATION"

The party system has been the basic organizational framework for American air accident investigators since Congress charged them with that responsibility. CAB investigators used it, and when CAB's Bureau of Aviation Safety was transferred to the NTSB, it kept the system intact. This type of investigative system allows those most financially vested in the outcome of the investigation to participate in finding out what caused the disaster.

The party begins like most other parties. The host and guests gather at a given location where the party is to be held. When they arrive, they meet and greet each other. Most of the guests know each other and inquire about each other's jobs, families, etc. The host will give them a tour of the site, an opportunity to ask questions, take notes, etc. After that, if nothing more needs to be done that day, the host and guests will gather for food and drinks, discuss the day's events, and reminisce about old times.

However, before the party really kicks off, the host—in this case, IIC Thomas Saunders—must, according to his superiors, lay down the rules that govern the event. The remarks are not off-the-cuff or improvised but written out by the Bureau of Air Safety verbatim so there is no ambiguity regarding the guests' roles during the party. At the first organizational meeting, the IIC must introduce himself and explain that he is organizing an investigative team. He explains, "The purpose of permitting the participation of organizations is not to enhance the position of these parties, but to assist the board in developing a more complete factual record."[1]

The host then warns that once an individual is designated as a representative of a party to the investigation, he must adhere to the following

19. "I did not participate in the NTSB's investigation"

rules in regards to the sharing of information found during the investigation:

1. No one will withhold information.
2. All information obtained by group members will be brought to the attention of the respective group chairman.
3. All information and developments ascertained during the investigation by the various groups will be passed to the IIC by the group chairman.
4. Each participating party will designate a coordinator for their organization.
5. Group members may pass factual information to their respective coordinators after this information has been made known to their group chairman.[2]

However, the rules of the information flow become stricter if information needs to be shared with those outside the party circle. The IIC makes known to all that

> all of the factual information and developments of the investigation that are made known to the Investigator-in-Charge will be passed on to each of the coordinators [sic] may relay information to their respective organizations provided the information is factual and in its right perspective. This information should be transmitted on a "need to know" basis for purposes of prevention, remedial action, or other similar reasons. The coordinators will keep the Investigator-in-Charge apprised of information so relayed. Common sense and good judgment should predominate in this matter. Do not discuss the investigation in public. At all times you must exercise caution when you discuss aspects of the investigation, as there may be unauthorized persons about you who might misconstrue or misuse this information, thus creating an adverse situation.[3]

Prater Hogue had an invitation to Saunders' parties, but only whenever a Boeing jet crashed. Hogue received the invitation because he represented, and worked for, Boeing during crash investigations. Tom Saunders and Hogue had worked closely during the fall of 1965 when the 727 had trouble staying in the sky.[4]

Tom Saunders knew most of Piedmont's employees through his brother. Some he had known for many years. Of course, if the party involved Piedmont, Tom's brother Zeke would show up and make the Piedmont assignments to the different committees. No one questioned this impropriety. After all, Bobbie Allen, Tom's boss, went on record and said that "the investigation cannot be subverted."[5] Besides, Zeke said he had not spoken to his brother about the crash. Under oath, Zeke Saunders stated,

"Upon learning the next day [July 20, 1967] that my brother Thomas Saunders was the National Transportation Safety Board's Investigator-In-Charge (IIC) of this accident investigation, I turned over my responsibilities regarding the accident investigation to other Piedmont employees. From that point forward, I did not participate in the NTSB's investigation."[6]

He was, however, still the boss of those whom he appointed to his brother's investigation, and he had access to those participants as long as they remained employed by Piedmont Airlines. Since the NTSB rules, though, state that a participant in an investigation cannot pass information other than facts to others within that individual's organization, and then only for the purposes of prevention, remedial action, or other similar reasons, Zeke Saunders and Piedmont Airlines should not have any more access to the causes of the crash than anyone not participating in the investigation.

The NTSB, like the CAB before it, promotes the party system because it surrounds its investigators with those people who are most acquainted with the technical expertise of the aircraft involved. According to the rules of the Bureau of Air Safety, "All persons participating in this investigation must be in a position to contribute specific factual information or skill which would not otherwise be supplied."[7] Then comes a warning about failing to follow NTSB rules. Participants are advised that if they fail to follow the rules, they may be removed from the investigative team and their organization may be removed from party status.[8]

The NTSB trumpets the party system because, theoretically, it believes that it ensures that all information regarding the crash will be brought forth among the group and discussed so that nothing is left out or hidden. Since the NTSB is one of the smaller agencies within the United States government, the party process allows it to draw on others' expertise because of the limited amount of NTSB staff available to conduct major investigations. The NTSB also likes their help because if a problem is discovered during the investigation with a certain party's product, that participant will have instant access to that information to fix the problem quickly if it is deemed a cause of the accident.

Opponents of the party system point out the inequity of representation within an investigation. Those closest to the investigation, victims of the crash itself as well as families, are not permitted to participate in an investigation. No members of the news media, lawyers, or insurance personnel are allowed to participate. A RAND study released in 2000 explained the dangers of the party system: "Parties that face potentially enormous economic losses if they are found to be the cause of an accident could attempt

19. "I did not participate in the NTSB's investigation" 159

to disrupt or bias an investigation."⁹ The party system allows those who may have caused the accident to play a part in the investigation. It is the same as asking a burglar to investigate a burglary he committed. It stretches the imagination to think that participants within an investigation will not have one eye on protecting their company during the investigation or sharing any information with their company. This is especially true if the one person asking for information holds the member's job in his hands, not unlike the relationship between Zeke Saunders and those he appointed from his company to participate in his brother's investigation.

In violation of the rules governing their participation in the investigation, certain Piedmont party members kept Zeke Saunders fully aware of every aspect of the investigation.

Just a week after the accident and a few days before Zeke Saunders assured Jack Anderson that he was not participating in the investigation, Saunders peppered John Reagan with questions about the flight data recorder. Saunders had assigned Reagan to the flight recorder group, and Saunders also knew that Reagan had been a part of that group's investigative efforts within hours of the crash. If anyone knew anything about the flight recorders onboard Flight 22, Reagan was that person. From newspaper reports and employee information, Saunders knew that Schulte had turned his aircraft at 4,200 feet, 800 feet below the FAA's published departure procedure, and it worried him. He needed more information.

O.E. Patton had advised Reagan before he left Washington, D.C., that he would receive a copy of the flight recorder report. On July 28, 1967, as promised by the NTSB, Reagan received the report at his Piedmont Airlines office in Winston-Salem. He read the report and then, disregarding the rules governing party status, walked down the hall and personally handed it over to Zeke Saunders.

"This was a report that I received from the National Transportation Safety Board covering the flight data recorder which was installed on Aircraft 650 at the time of the accident," Reagan stated as he gave his boss the document.[10]

Saunders read the report while Reagan waited patiently for his boss to finish. When he had finished reading, Saunders asked Reagan for an interpretation of the "G" forces as indicated on the top of the page of the graph and asked what caused it.

Reagan speculated that it "could have been turbulence caused by cloud cover, clear air turbulence," but that he had no way of knowing for sure.[11] Clearly, Saunders did not want factual information; he wanted interpretations—a clear violation of party status rules according to the NTSB.

But it just was not Zeke Saunders looking for information beyond facts from his party representatives; director of flight operations Tadlock also wanted information, but he was a little more circumspect. He attempted to read out the heading trace on the report and then kept asking Reagan if he was correct in his interpretation.[12] Of course, all this information would be gold for the lawyers due to represent Piedmont in the crash and would certainly give them a huge advantage if the case went to trial.

Zeke Saunders spoke more freely with Frank Nicholson, who had been assigned to the ATC group. They had known each other for years, and both were veterans of Piedmont. Nicholson had also known Tom Saunders for over twenty years. Nicholson would share anything he could with his boss.

In complete violation of the NTSB rules governing the party system, Saunders and Nicholson discussed anything and everything regarding the crash. In litigation testimony several years after the crash, Nicholson admitted as much. When asked if he had any discussions with Mr. Saunders concerning the crash, he answered affirmatively. "Yes, sir, we had many discussions and conferences."[13]

When asked what they discussed, he answered generally. "I don't really recall just what we did talk about. I suppose we talked about just anything that was involved."[14] As he freely admitted, Nicholson did not observe any boundaries in his conversations with Saunders. Everything he knew about the crash was discussed with Zeke Saunders. Nicholson had information regarding the Cessna's flight path and the tapes from the cockpit voice recorder, and he also had the statements from the controllers in Asheville and the people in Atlanta Center.[15] Frank Nicholson had virtually all the information regarding the crash and freely shared it with Zeke Saunders.

But Zeke Saunders' information did not come only from his own employees. His brother's friend from Boeing, Prater Hogue, personally forwarded information from Boeing's analysis of the flight recorder to him.

Prater Hogue had been with Boeing for almost thirty years. His familiarity with all Boeing aircraft, military as well as civilian, made him Boeing's representative at crash sites involving Boeing jets. He had worked closely with Tom Saunders during the spate of accidents involving the 727 in 1965–1966. He was also intimately aware of the special study of the 727 and its handling capabilities. His main job consisted of making sure Boeing aircraft were not at fault for accidents. Boeing paid his salary to represent their interests.

Soon after the accident, the NTSB asked Prater Hogue to have Boeing

19. "I did not participate in the NTSB's investigation" 161

read out the data on 68650's flight recorder. In a letter dated August 8, 1967, file number 6–7/11–1334-I/C, Prater Hogue returned the results of Boeing's findings to Zeke Saunders at Piedmont.[16] The only explanation available of why Hogue felt the need to share Boeing's analysis with Zeke Saunders at Piedmont was that someone at the NTSB requested him to do so. While it will never be known for certain who authorized the information's transmittal, Tom Saunders must be ranked at the top of the list.

Clearly, Piedmont personnel did not follow the rules regarding the dissemination of information during this NTSB investigation. Without a doubt, the actions of Reagan, Nicholson, and Hogue made a travesty of the party system, and Piedmont should have lost its party status because of it. If these acts were committed with the consent of Tom Saunders, he should have been removed from the investigation for his inability to maintain impartiality. If the acts were committed without his consent, he had completely lost control of his investigation.

20

"Since we are in a real budget squeeze, aviation must take its lumps"

Secretary of Transportation Alan Boyd had listened to the infighting within the administration for weeks and knew the time had arrived for the president to make a decision on aviation. If he did not lead on the aviation issue, Congress would force him to act, and Lyndon Johnson did not like anyone telling him what to do. The House of Representatives had treated McKee brutally. It had been some years since any representative of any administration had been accused of dereliction of duty. Boyd knew McKee would not take that lying down. He had witnessed McKee's battles with the director of budget, Charles Schultze, in trying to haggle for more money for the FAA. Boyd also sympathized with Schultze because Johnson pushed him to fund both the Vietnam War and the Great Society. Boyd, however, did know one thing: If another midair collision occurred, it could possibly cost Lyndon Johnson the 1968 presidential election.

William McKee had had enough of the budget games the administration played. He had pleaded with both Johnson and Schultze for more funds to hire more air traffic controllers, but their resistance had proved too much. The war in Vietnam had priority. McKee had gone to Congress and gamely argued the administration's stance on the tight budget and had been made to look like a fool. He needed an outlet to tell his story and found it in one of America's most influential newspapers. Charles Schultze may well have choked as he read the *New York Times* on the morning of September 18, 1967.

McKee had gone to the *Times* and revealed the battle he and Schultze were waging. He framed the argument for more money as an issue of safety. He revealed to reporter Evert Clark that the FAA could not keep up with the air system with the currently allocated funds. Air safety was compromised. McKee revealed to the reporter that he had asked for $100 million to fund the inadequacies at the FAA. He loved the next paragraph. It read that "if a denial of the request were to be followed by a major accident involving air traffic control, the safety issue could be catapulted into a political controversy."[1]

The article further stated that the air traffic control system had increased in size by 178 percent since 1963 "without the hiring of any appreciable number of new controllers."[2] Knowing that the congressmen who treated him so shabbily at the last hearing would be reading the *Times* article, he made sure to secure a great quote: "I am concerned about the number of personnel, particularly in our air traffic control system."[3] Satisfied that the battle with Schultze now had received a public airing, he waited for the White House to respond to his challenge.

Like John McNaughton, the late secretary-designate of the navy, Charles Schultze came to Washington at the behest of John F. Kennedy and stayed to serve Lyndon Johnson. He had received two degrees in economics from Georgetown University, and the University of Maryland awarded him a PhD in economics in 1960. He had taught economics at Indiana University before Kennedy asked him to come to Washington.[4]

While Schultze taught college students, William McKee played the Washington political game. But by 1967, Schultze had also become a Washington player. Even he knew that McKee's argument for more funds on the grounds of air safety could not go unheeded. McKee had won this turf war using the single word "safety," and it was Schultze's unenviable responsibility to advise the president that McKee had won this battle and to fork over the money.

On the same day as the *Times* article, Schultze wrote Johnson that by "arguing his case on safety grounds, McKee is trying to put the monkey on your back, Boyd's back and mine."[5] Schultze argued that this increase in funds "is *not* a question of safety, but one of funds for the *convenience* of private pilots and commercial air travelers."[6] Schultze's maintained that safety in the air could be achieved within any reasonable budget. He argued that all McKee had to do to achieve air safety was to increase the "spacing between aircraft in flight; reduce the flow of traffic more quickly when bad weather sets in; and add to the time between aircraft arriving and departing at airports during peak periods of traffic."[7] By implementing these regulations, Schultze argued to the president, air safety would be increased. He admitted in his memo that air travel might be delayed by enforcement of these regulations, but "they can maintain safety."[8] Before offering his solution to the problem, Schultze complained once more about McKee's tactics. "Since we are in a real budget squeeze, aviation must take its lumps along with space, education, and all the rest—even if this does mean greater delays."[9]

Schultze's solution came straight out of a magician's hat. He advised the president to send to the Senate a *"budget amendment* which increases (by $7 million) the funds for FAA operations to make the most urgent improvements."[10] Schultze wanted this money taken from the FAA's unobligated equipment appropriations so that no additional funds would be used. Schultze argued, "It represents a quick fix to meet the current situation, but recognizes our current budgetary problem."[11]

He also advised Johnson to send a letter to Boyd instructing him to direct McKee to *"institute whatever flight regulations are necessary to maintain safety within the budget the Congress provides*—this puts the monkey right back on McKee, where it belongs."[12] Schultze also advised Johnson to tell Boyd to tell McKee to "propose a comprehensive long range plan for FAA's air navigation system and companion legislation to levy sufficient user charges to pay for the improvements, to keep the load off the general taxpayer."[13] As an aside, Schultze added, "The attached *New York Times* story this morning indicates the kind of games that FAA is playing."[14] McKee may have gotten his money, but Schultze was going to try to embarrass him in front of the president every chance he got.

By the fall of 1967, Lyndon Johnson's presidency was sinking, and he grasped at every life preserver thrown to him. The Vietnam War was getting worse by the day, race riots rocked the streets of the nation's cities, and Congress had stopped rubber-stamping his proposals after the 1966 election. His credibility was at an all-time low, and the network news stations skewered him and his handling of the nation's problems almost every night. Even aviation was in tatters. While discussing aviation issues with the president, Alan Boyd bluntly told him, "It is clear, however, that another major aircraft accident this year, particularly a mid-air collision, will cause the roof to fall in on you, McKee, and me."[15] Johnson had to act quickly.

Schultze offered Johnson an ideal solution to his thorny aviation problem. Johnson astutely recognized, and he had been warned, that if another midair collision occurred without an increase in the budget for the FAA, the nation's travel habits might come to a standstill. No one would want to fly, and it would take months to reassure the public that the nation's airways were safe. That would be a hard enough sell by itself, but with his inability to reassure the nation all was going well in Vietnam, it would be almost impossible for Lyndon Johnson to do it. By cloaking his department's budget requests in terms of safety, McKee had effectively trumped anything Schultze, or the president, could throw at him. Johnson agreed with Schultze that he could give the FAA more funds without increasing its already authorized budget and that it meant he could hire about nine hundred more people for the FAA.[16] On the evening of September 19, 1967, at 6:00 p.m. two months to the day after the crash of Piedmont Flight 22, President Johnson approved the hiring of nine hundred additional people for the FAA to better control the nation's airways. Eight hundred would be assigned to the air traffic control system and the other one hundred to flight standards.[17]

The next day, Johnson released his directive to Alan Boyd. Johnson stated that the rapid growth of aviation was "creating demands for substantial expansion and improvement in the Nation's air traffic control system. The federal government is the manager of this system."[18] Johnson admitted that federal funds must be spent to maintain the system. He also wanted the users of this system to pay their fair share. "I do not believe that the general taxpayer should be asked to shoulder this burden."[19] A user fee plan, though, would have to be passed through Congress, and that would take time and there was no time to waste. "Looking towards the immediate future, I am today submitting to the Congress a budgetary amendment designed to provide a more effective use of Federal Aviation Administration funds in the operation of the air traffic control system."[20] The president also took Schultze's advice and demanded that McKee review regulations to make the system safer, and that if delays were incurred, that would be a necessary evil. Johnson ended his letter, expressing pride in his staff. "This nation has an enviable record of air safety. I know that you and your associates can maintain this record."[21]

The midair collision of Piedmont Flight 22 became the catalyst for the hiring of more air traffic controllers. Aghast at the deaths of eighty-two people on July 19, 1967, a handful of congressmen and union officials courageously spoke out against William McKee's and President Johnson's cost-cutting measures and forced Johnson to authorize the hiring of almost one thousand additional personnel to secure the nation's airways. Now he, and everyone affected by the tragedy, waited for the NTSB to decide what had caused this senseless accident.

21

"WE, THE PUBLIC, PREFER SIMPLE ANSWERS"

It is at the discretion of the chairman of the NTSB whether or not to hold a public hearing in the case of a particular accident. Public hearings are touted as an opportunity for the NTSB to publicly present the facts of an accident to those interested. Of course, those most interested in the case would be the families of those lost in the accident. Close seconds would be the aircraft manufacturer, airline involved, and other entities associated with the performance of the flight: in other words, those with the highest financial stake in the matter.

The function and purpose of a public hearing evolved as each year passed, with its definition expanding or constricting depending upon the circumstances of a crash. The importance of a public hearing to the outcome of an investigation is so minimal that many times, public hearings are not even held. By the time a public hearing is held, the on-scene investigation is complete, witness statements have been taken, wreckage has been analyzed, and group chairmen and their investigators have already completed their reports. It is no surprise that by the time a hearing is held, investigators generally know what caused the crash.

Public hearings generally do not garner a lot of interest from the public. The only persons who show up at hearings are attorneys, family members of the deceased, and reporters. Most of what is discussed at a public hearing can be read about in the paper. Transcripts are taken and then sold for a fee from the transcribing company. What transpires at a public hearing is generally forgotten, remembered only in old newspaper stories written by reporters who do not have an understanding of aviation. The main product of the investigation, the only part of it that the attorneys, manufacturers, and airlines are concerned with, is the final report, and what is discussed

21. "We, the public, prefer simple answers" 167

at a public hearing usually does not make it into the final report. That is because what is discussed is sometimes so disjointed and confusing that one cannot help but wonder how the skies remain as safe as they do.

Public hearings have been a part of the investigative process since the 1940s. However, the participation of parties to the investigation during the public hearing has not. Since the early 1940s, only members of the Air Safety Board could question witnesses. In 1956 that changed; the CAB allowed parties to the investigation this privilege also.[1] Again, air carriers, manufacturers, and other parties whose equipment or procedures, or personnel may have caused the accident were allowed to participate in another facet of the investigation. Family members, though, and attorneys representing them were not allowed to participate and question witnesses.

In an effort to excuse this inequity, the CAB explained that the purpose of allowing them was "not to enhance the position of these parties but to assist the board in developing a more complete factual record."[2] It also had a safeguard measure to ensure that the rules would be followed. Simply, it would be up to the hearing officer to make sure of that. Francis McAdams, one of many men who eighteen months earlier had cleared the Boeing 727 when its safety was questioned, sat as the chairman in charge of this particular hearing.[3]

The three-day public hearing opened on Monday, October 9, 1967, in the ballroom at the Grove Park Inn in Asheville, North Carolina. As the hearing opened, even Chairman McAdams seemed confused about the function of public hearings. During one discussion about the purpose of a public hearing, McAdams said, "Apparently, there is some misunderstanding as to the purpose of the hearing. This hearing is a part of the investigation. It is an integral part of the preliminary fact-finding process, and that's all it is."[4] A few minutes later, McAdams contradicted himself. In explaining the role of the NTSB, McAdams stated, "As part of the Board's responsibilities, it investigates accidents. In some accidents it holds a public hearing to assemble facts. There is no need for the board to hold public hearings to assemble facts. In fact, the great majority of the Board's determinations of probable cause are based upon facts that were not adduced in a hearing of this type."[5]

One of the main issues raised at the hearing was whether the crew members of the Piedmont jet violated departure procedures by turning left before they reached five thousand feet as the departure procedure indicates. The AOPA had indicated that the crew of the 727 did violate the departure procedure by turning before five thousand feet. The FAA countered that the procedure was not violated, saying that the procedure was merely a "device to assist" pilots in avoiding the mountainous terrain near the airport.[6] Captain Paul Snell of Piedmont Airlines testified that a pilot would not be in violation of the departure procedure if his aircraft turned under five thousand feet. However, when asked if he would follow an air traffic controller's command that he turn before five thousand feet, he replied, "I might question it when I receive it."[7] He then quickly added that ATC could only change the procedure in the case of an emergency.[8] It quickly became clear that members of the aviation community could not agree whether published departure procedures at the Asheville airport were mandatory or advisory. It would be up to the NTSB to make that decision in its final report.

Another important matter brought up at the hearing was the notification by controller James Watkins to the pilot of the Cessna regarding his landing instructions and to the pilot of the Boeing regarding his departure procedure. Transcripts showed that Watkins, at first, advised Addison to proceed to the Broad River radio beacon and then changed his mind and told him to proceed to the Asheville radio beacon. It is clear that Watkins did not tell Addison what type of approach to make and only did so about a minute before the collision. It is also clear that Watkins advised Schulte to turn prior to reaching five thousand feet.

Watkins did not accept any accusation that his actions might have contributed to the accident. He even stated that had both pilots followed his instructions, the accident would not have occurred.[9] According to a newspaper account of the hearing, "Watkins made a point of telling the panel that he had notified the Cessna pilot seven minutes before the crash to swing north of the airport."[10] Nowhere did Watkins admit in his response that he had, in fact, told Addison to proceed to the Broad River radio beacon first. Nor did he admit that he did not tell Addison what type of landing to use. It would be up to the NTSB to make a determination of Watkins' actions in its final report.

The final major issue dealt with at the hearing concerned the cigarette fire in the cockpit of the Piedmont jet. During the first day of the hearing, the transcript of the CVR on the Piedmont jet was read into the record of the hearing. Nothing was left out of it. The main question asked of Piedmont pilots at the hearing was whether the crew was authorized to smoke just after takeoff.

Paul Snell responded by saying he had never seen any regulation that stated a crew could not smoke during the flight phase of an aircraft operation.[11] However, Captain Tadlock, director of flight operations for Piedmont, stated that Piedmont's written policy in its Operations Manual states the crew "shall not smoke in the cockpit if the cockpit door is open and where it would be visible to passengers."[12] He also stated that the general practice of Piedmont crews was that "they don't smoke when the passengers are not allowed to smoke"[13] and conceded that from the time Captain Schulte lit his cigarette to the time of the collision, the No Smoking sign was on for the passengers.[14]

The question of whether or not a crewman could smoke during flight, or when the No Smoking sign was on, was really not the issue. The issue was whether cigarette smoking, or allowing a cigarette to catch fire and distract the crew, was proper operating procedure in a Boeing 727. These men who sat in judgment at the hearing, the men like Francis McAdams, Bobbie Allen, and Tom Saunders, all agreed eighteen months earlier that the 727 "*does have* highly responsive and versatile flight characteristics and that these favorable characteristics may be misleading to the pilot, or are presenting the impression that greater liberties may be taken with the aircraft in normal operating situations...."[15]

These same men, the men who cleared the 727 in earlier accidents, had written: "The Board must re-emphasize that the responsibility and authority committed to an airline captain requires the exercise of sound judgment and strict adherence to prescribed practices and procedures. Any deviation can only result in a compromise of aviation safety. Airline management, too, has a heavy responsibility for devising, developing, and implementing methods of procedures designed to insure that all of their pilot personnel constantly exercise a conservative, prudent, approach to their daily work."[16] Yet when the time came to reemphasize the special care that must be used in flying a 727, these very men, one of whom was the brother of a Piedmont vice president, remained silent. Surely, they would address the issue in their final report.

The political and social landscape of the United States changed completely in the eleven months between the end of the public hearing and the release of the final report on September 5, 1968. Interest in the NTSB's findings had not so much waned as it had been replaced by history-making events. After years of denial that America could be beaten in the Vietnam War, and the continued pronouncements that the Communists were beaten, on January 31, 1968, North Vietnam launched the Tet offensive. In the largest campaign of the war up to that point by either side, Communist troops conducted a countrywide offensive, infiltrating the very infrastructure of the American military presence. It stunned all of America but none more so than Lyndon Johnson. Two months later, on March 31, 1968, he announced that he would not seek another term as president, effectively resigning from the presidency with a January 21, 1969, departure date.

Four days later, on April 4, 1968, the civil rights movement itself was attacked when an assassin killed Martin Luther King, Jr. in Memphis, Tennessee. Five days of riots consumed the country, and in Washington, D.C., rioters came within two blocks of the White House as the government mounted machine guns on top of the Capitol to keep them back.

Just over two months later, while running for the presidency, Robert F. Kennedy, brother of Johnson's predecessor, was shot and killed by an assassin. More riots followed, and the Democratic presidential primary process was thrown into chaos, with Johnson's vice president, Hubert Humphrey, ultimately prevailing at the Democratic convention and receiving his party's nomination as president.

On November 5, 1968, the American people sent Richard Nixon to the White House. Nixon, who lost to John F. Kennedy in 1960, had retired from politics in 1962 but saw an opening in the dissatisfaction with the Johnson administration. With his election, the careers of those involved with Johnson were effectively over. McNamara, Boyd, McKee—all would be gone when the new administration came to power.

With these tumultuous events convulsing the country, there is no doubt that the release of the final report of an eighteen-month-old plane crash would be met with little interest. Only those with a financial or emotional stake in the results of the report would even glance at the findings. Too much had happened in the country since the planes collided to garner the accident much attention any more.

The NTSB released the report on the Flight 22 midair collision on September 5, 1968. The probable cause read as follows: "The Safety Board determines that the probable cause of this accident was the deviation of the Cessna from its IFR clearance resulting in a flight path into airspace allocated to the Piedmont Boeing 727. The reason for such deviation cannot be specifically or positively identified. The minimum control procedures utilized by the FAA in the handling of the Cessna were a contributing factor."[17] The Safety Board determined that Piedmont Airlines bore absolutely no responsibility for the accident.

The report stunned William McKee. The FAA had never been mentioned as the cause of an accident in a probable cause statement. As Francis McAdams stated, "As I have mentioned previously the FAA, insofar as I know, has never been cited as being a probable cause of an accident."[18] He added "the report is the strongest comment with respect to the deficiencies in FAA procedures that has been released by the Board or its predecessors."[19] McKee could take some comfort—being cited as a contributing factor is less severe than being cited as the probable cause.

The probable cause statement, mandated by Congress on the NTSB and its predecessor the CAB, has always stirred controversy within the aviation community and within the NTSB itself. The probable cause statement limits the Safety Board to find the one factor that caused the accident. Francis McAdams wrote that the definition of probable cause is "the most reasonable cause, the one having more evidence for than against."[20] This, however, greatly limits the scope of an investigation, especially if there are multiple factors involved. C.O. Miller, who succeeded Bobbie Allen in the summer of 1968 when Allen resigned as the head of the Bureau of Air Safety, was a great critic of the probable cause statement in accident investigation. Just one month after the final report was released, Mr. Miller gave a speech in California. Obviously referring to the Piedmont accident as an example, he said "There is one fundamental overriding principle in safety that requires a diminution of emphasis on a singular probable cause. Accidents are a combination of events with numerous chains or cause-effect relationships determinable therein."[21] He felt that the reason Congress limited the NTSB to finding a probable cause to an accident was because "we, the public, prefer simple answers; we like a nice, convenient, singular cause-effect relationship. It avoids adding another confusing segment to our already complex lives."[22] The Piedmont accident report had its first critic.

With Bobbie Allen's departure from the Bureau of Aviation Safety, Miller inherited the open cases on which the board had not submitted a final report. From the moment he looked at it, Miller could see that the Piedmont case could not be explained by a single probable cause. That alone would not explain why the two planes collided. Miller had to completely change the thinking of crash investigators, because to his amazement, he soon realized there were many on the staff within the Bureau of Aviation Safety who could not even explain the difference between the probable cause of an accident as compared to other types of causes with lesser degrees of liability. Miller quickly surveyed his group of investigators, both verbally and in writing, and came up with some startling results:

1. Lack of clear understanding of probable vs. prox. [proximate cause].
2. Desire for more uniform policy.
3. Acknowledged tendency to structure report towards singular cause.[23]

Confining themselves to a probable cause statement in the Hendersonville accident, the investigators of this case allowed other varying degrees of liability to be ignored, thus allowing many of those responsible for the accident to escape blame and the financial consequences that go with it.

Miller rapidly began to downplay the singular probable-cause statements. Chairman O'Connell soon saw the benefits of Miller's thinking. In a speech to safety technicians, O'Connell said, "We must not become so preoccupied, in sifting through the wreckage and seeking out that one alternative of omission or commission that caused the event, that we ignore other lessons which may be lying on the surface.... We place a great premium on pursuing the many alternatives, narrowing down the areas of consideration, and arriving at that one specific cause. I have been wondering whether it is so important that our choice of the cause be so exact.... Is it always vital that we identify beyond peradventure which one was the culprit? I think not. If the alternatives are all possible causes, shouldn't we seek remedies for all? I think so."[24]

Miller also had another reason for diminishing the probable cause statement. "Implicit also is a concern," he wrote, "that if the Bureau is ever exclusively identified with *a* probable cause determination, then its product will tend to be vectored away from safety promotion in the minds of the public and will be identified with fault-finding, pure enforcement, or legal functions."[25] Miller wanted the bureau to become a key player in promoting all aspects of safety, rather than just finding one cause to an accident.

Francis McAdams, though, remained comfortable with a probable cause statement. His definition of a probable cause was "the most reasonable cause, the one having more evidence for than against."[26] He also believed that it could be expanded to include contributory factors, but they would not carry as much weight as a probable cause or have larger consequences than a probable cause.[27] Even limiting themselves to a probable cause statement, the NTSB could have cited Piedmont with at least two contributing factors: failure to keep an appropriate lookout and lack of cockpit discipline. However, they chose not to, and cockpit safety and resource management was pushed back seven years because of it.

The final report submitted by the NTSB in the case of Piedmont Airlines Flight 22 and Lanseair Cessna 310 was not an accurate reflection of the accident. Prepared by a team of investigators led by the brother of the vice president of Piedmont Airlines, it exonerated Piedmont from any blame in the accident. Every time an opportunity arose to cite Piedmont, the NTSB investigators chose not to do so. Every time an opportunity arose to blame the Cessna, the NTSB investigators chose to do so.

The investigators gathered the information from the weather bureau at Asheville airport and used this as their basis for explaining the weather conditions at the time: "The ceiling was estimated 2,500 feet broken. No local pilot reports were available.... The visibility was four (4) miles in haze and determined by visibility markers as shown on attached visibility charts."[28] Nowhere in the final NTSB report does it state that the rotating ceilometer at Asheville airport, a device used to measure the height of clouds in the area, was undergoing maintenance at the time of the collision and was not in use even though the meteorologist in Asheville stated it in his report.[29] The only true measure of the visibility in the area would be from pilots. Of all the pilots in the area, the one whose recollection should be given the most weight would be J.J. Barnes, the pilot who saw the accident from one mile away and immediately called the accident into the tower. In his witness statement, he wrote, "I would say weather conditions at the time of the incident were broken with the tops at approximately 7,500 feet. There were some towering cumulus; however, not in the immediate vicinity of the incident. There was existing at the time of the incident a Southwest, Northeast trough which appeared three (3) or four (4) miles wide and was the location of the incident."[30] Clearly, according to the one man in the sky who witnessed the incident, the area of the accident was free of clouds. The NTSB, though, ignored Mr. Barnes's observations. The report states, "Although witnesses reported that the collision occurred in an area clear of clouds, the evidence indicates that both aircraft would have been operating in and out of broken clouds just prior to the accident. Therefore, in this situation, the 'see and be seen' concept can only be considered inapplicable."[31] The wording of this description is odd: "would have" rather than "were."

The NTSB painted clouds in the vicinity of the accident in its own imagination. No basis or evidence exists for believing that clouds obscured the view of either pilot before the collision. In their own group report led by NTSB investigator Russell Abbott, NTSB investigators interviewed 134 persons and obtained 80 written statements from them. Out of these witnesses, only five reported cloudy conditions in the sky that day.[32] Clearly, the NTSB ignored its own data when coming to the ribald conclusion that both aircraft were operating in and out of clouds.

As for the operation and maintenance of the Boeing 727, the NTSB chose to ignore that the onboard radar unit did not work, a radar unit that personnel from the Boeing company said could see other aircraft and therefore function as a primitive collision-avoidance system. They dismissed the pilots' complaints regarding the unit with a simple nod to expediency. Investigators wrote that "both aircraft were properly certificated and maintained in accordance with existing requirements."[33] They also failed to mention anything regarding the inefficiency of FAA oversight of Piedmont Airlines and its Boeing 727 aircraft. The NTSB failed to see the importance of having safety inspectors with oversight authority at Piedmont Airlines' headquarters in Winston-Salem—inspectors who were certificated in the one jet aircraft the airlines' pilots was flying. It went without mention.

Also going without mention were the performance inadequacies noted during Captain Ray Schulte's training and line checks. While the NTSB did acknowledge that Schulte had received an unsatisfactory grade during his rate check, it made no mention of his "very poor" altitude and airspeed control during his initial training. The report made no mention of the failed line checks; his weaknesses in positive command, using the flight director system, and setting the altitude reminder; or his "ragged" performance in IFR procedures and trip management. None of this was mentioned in the report. Regarding the line checks, the final report stated of Schulte that "he passed his last line check in the B-727 on July 6, 1967."[34] When taken in the context of the dangers of the Boeing 727 aircraft and the warnings of the CAB regarding the operation of that aircraft, Piedmont Airlines failed in its responsibility to place its best pilots as captains in command on that aircraft; it failed to have a safety program in place to properly train pilots that they must, at all times, exercise sound judgment while flying that aircraft. In addition to Piedmont's omission of this critical information, the NTSB failed in its assignment to identify and publicize gaps in training and safety procedures so that those errors would not be repeated.

The most controversial statement in the report had to do with the actions of the Piedmont crew. It is known that Captain Schulte, in defiance of the Bureau of Air Safety's comments regarding the safe operation of a Boeing 727, smoked a cigarette during the busiest part of flight and the cigarette caught on fire, distracting the crew. After Schulte extinguished the cigarette, the crew bantered back and forth about that night's upcoming steak dinner. This was during a time in which, had the crew practiced proper cockpit vigilance, they might have been able to see David Addison's Cessna 310 bearing down on them from the front left side—Captain Schulte's side of the cockpit. In its report, the NTSB admitted that each crew had thirty-five seconds in which to view each other.[35] Thirty-five seconds is more than double the time needed to react to the presence of another aircraft.

NASA studies on reaction time are based on: see the object, recognize it as an airplane, perceive a collision course, decision to turn left or right, muscular reaction, and airplane response time. Those steps add up to 12.5 seconds cumulatively.[36] The only way to exonerate the actions of the Pied-

21. "We, the public, prefer simple answers"   175

mont crew would be to ignore the conversation recorded on the CVR, and that is exactly what the NTSB did. The report stated, "The conversations recorded on the tape were concerned primarily with the operation of the aircraft and nothing was found of a probative value to the investigation."[37] In the eyes of the NTSB, the Piedmont crew acted in accordance with prescribed policies and procedures rather than in defiance of their own recommendations on the Boeing 727 jet.

The NTSB also exonerated the Piedmont crew for turning at 4,200 rather than the 5,000 feet published by the FAA in its departure procedure manual. Ignoring the comments and concerns of many aviation professionals, the NTSB determined that an aircraft did not have to follow published FAA procedures when departing Asheville airport. "The flight of PAI [Piedmont] 22 was brief and involved only the takeoff and climb out to the point where the collision took place,"[38] the report stated. "The conduct of the flight was in accordance with its IFR clearance and within the confines of the applicable procedures and regulations. The Board concurs in the FAA interpretation that the applicable IFR departure procedures were established for terrain-obstruction avoidance purposes and are not mandatory procedures when a departing flight can effect terrain avoidance by visual means."[39]

If, as the investigators wrote in their report, "both aircraft would have been operating in and out of broken clouds just prior to the accident," how could the pilot effect terrain clearance by visual means? The NTSB cannot have it both ways; investigators cannot clear the Piedmont crew for not seeing the Cessna due to broken clouds and then exonerate them for being able to see the mountains due to clear weather. This just reinforces the notion that the NTSB may have manipulated its weather data to clear the Piedmont pilots. But the IFR departure procedures were not established for terrain-avoidance purposes. Certainly, the NTSB had the memo from Earl Cato and the departure procedures that went into effect for Asheville in 1961. This was the memo that stated the departure procedure was established to expedite air traffic, not for any terrain avoidance. If they read it, they ignored it, just like the conversation about the cigarette fire.

James Watkins fared a little worse than Piedmont Airlines. His actions in the tower while handling the Cessna led to the FAA, for the first time, bearing some of the blame in an air accident. Investigators reproached Watkins for not notifying Addison, at the time of first radio contact, or as soon as possible thereafter, the type of approach to use in Asheville. They wrote, "Section 262.7 of AT P7110.1B provides that Approach Control facilities will notify an arriving aircraft at the time of first radio contact, or as soon as possible thereafter, the type approach clearance or the type of approach to be expected when two or more approaches are published and the clearance limit does not indicate which will be used. This was not done."[40]

They further stated that because Addison did not get that information, he would probably have proceeded on the basis of his last information from Atlanta control, which was to expect an ILS landing in Asheville. Compounding the error was Watkins' next transmission to Addison where, after telling him to head to Broad River, he corrected it and sent him to the Asheville radio beacon. The board wrote that at this time, Watkins knew that an ADF-2 landing was to be expected "since there is no ILS procedure utilizing the Asheville Radio Beacon."[41] Had Watkins told Addison to "plan for an ADF-2 approach at the time of first contact or at least when the clearance to the Asheville RBN was given, any confusion or misunderstanding by the pilot as to the approach to be conducted, or as to the location of the Asheville RBN, would have been eliminated prior to passing the VOR."[42] In other words, had Watkins followed ATC procedure, the accident would not have happened.

Perhaps to protect the FAA from public embarrassment, the Safety Board chose not to use harsher language in singling out Watkins' performance within the final report. An internal NTSB document, though, written by Francis McAdams within months of the report's release, indicated that the NTSB could certainly have been harsher on the FAA. So harsh, in fact, that they could have cited Watkins' actions as the probable cause of the accident. In a memo to Chairman O'Connell, McAdams stated, "The air traffic control system is a cooperative system and the FAA has assumed responsibility for the separation of aircraft. This is a joint responsibility equally divided between pilots and controllers."[43] He explained that while the pilots have greater knowledge of their aircraft, tower personnel have expert knowledge of the airport, its directional beacons, and the location of other aircraft. "The pilot under such a system must rely upon the superior knowledge of the controller with respect to other aircraft and the accuracy, adequacy, and timeliness of the clearances."[44]

Since radar was not available in Asheville, McAdams stated that "it was absolutely incumbent upon the controller to know with the highest degree of certainty the geographical location of the Cessna before clearing Piedmont into that airspace."[45] While leaving out pertinent information from the final report that would undoubtedly have placed the probable cause of this accident upon Watkins and his actions, McAdams used the information in his memo. "The controller cleared Piedmont based upon not only a garbled transmission, but a transmission all of which the controller admitted he did not hear. Further, the transmission was vague or ambiguous with respect to which Asheville facility the Cessna was proceeding to. More important, the controller failed to give the Cessna all of the available information in a timely fashion."[46] He said that the controller should have realized that the charts into Asheville could cause any pilot confusion and that it was his responsibility to clear up any confusion. McAdams finished with this admonition: "In conclusion, the controller performed his duties in a minimal fashion and this is not sufficient under the circumstances. At the very least, the controller should have requested a read-back from the Cessna."[47]

This memo provides strong evidence that Watkins acted poorly in his handling of Addison's flight. McAdams states that Watkins' handling of the flight led to Addison's confusion in following Watkins' commands, not the other way around. He admits that Watkins did not hear Addison's transmission yet continued him on a path right into the Piedmont jet. None of this information, which McAdams and Tom Saunders had in their possession, made it into the final report, which exonerated the FAA as *the* probable cause of the accident.

That leaves, according to the NTSB, David Addison to blame. However, all Addison did was follow the instructions from the air traffic controller handling his flight. Atlanta Center told Addison to expect an ILS landing at Asheville. While this was not a clearance, it certainly gave Addison time to get his charts ready and line up for a safe ILS landing. When Atlanta Center transferred Addison to Watkins, Watkins did not indicate that the approach had changed. As a matter of fact, Watkins told Addison to proceed to the Broad River radio beacon, which is part of the ILS approach into Asheville. Seconds later, Watkins changed his command and had Addison proceed to the Asheville radio beacon. Addison had no idea if the Broad River radio beacon and the Asheville radio beacon were one and the same. With seven facilities surrounding the Asheville airport having the name "Asheville," Addison could easily have been confused as to the location of the Asheville radio beacon, because it is not on an ILS approach.

As he flew over the Asheville VOR on Sugarloaf Mountain, Addison, according to the Bureau of Standards, told Watkins that he was heading off the Asheville VOR via the 238 radial. This should have clued Watkins, whom McAdams stated in his memo had superior knowledge of the airport, that Addison was not heading towards the Asheville radio beacon. Watkins admitted under oath that he did not hear the transmission because he was performing other duties. As McAdams stated, at the very least, "Watkins should have requested a read back."⁴⁸ How James Watkins' failure to follow ATC procedure became Addison's fault has never been adequately explained.

As for the "garbled" transmission from Addison in which he indicated his heading of 238 off the Sugarloaf VOR, Francis Graves ignored the Bureau of Standards report. That report was not even mentioned in the final report. A footnote was inserted into the final report. It stated, "The pause in the main transmission is approximately 4 seconds long. Background conversation is audible during this pause; however, despite extensive examination, no reliable intelligence could be determined."⁴⁹ That statement is far removed from the one Graves wrote in his group chairman report, which stated, "Considerable effort was expended to gain intelligence from that portion of the recording through electronic filtering of unwanted sounds from the tape. That effort did succeed in obtaining near complete intelligence."⁵⁰ Between September 12, 1967, the date of his report, and September 5, 1968, the date of the release of the final report, it seems that the "near complete intelligence" disappeared completely.

Four months after the release of the accident report, America got a new president in Richard Nixon. The war in Vietnam continued to rage, and protests gripped the country with a fervor stronger than in the years preceding. Nixon expanded the war into Cambodia and Laos, and American soldiers kept dying. Vietnam became the issue that consumed the American public in the late sixties and early seventies; almost nothing else mattered, certainly not a plane crash in the hill country of western North Carolina that claimed the lives of eighty-two people.

22

"Poor cockpit discipline"

Tragedy struck again in North Carolina in 1974, when Eastern Airlines Flight 212, a DC-9, crashed short of the Charlotte-Douglas airport. Seventy-four people lost their lives. While the pilot and copilot engaged in talk of politics and cars, the aircraft descended until it clipped some treetops and smashed into the ground. The NTSB determined that "the probable cause of the accident was the flight crew's lack of altitude awareness at critical points during the approach due to poor cockpit discipline in that the crew did not follow prescribed procedures."[1]

The Flight 212 accident began a series of safety recommendations that addressed professional conduct by pilots while operating their aircraft. The NTSB became very concerned by the lack of attention paid by crews in commercial aircraft, especially under ten thousand feet during the most critical phases of flight. Less than a month after the Flight 212 crash, the NTSB issued Safety Recommendation A-74-85, urging the FAA to begin reminding pilot associations and airline management to "form new, and regenerate old, professional standards committees to monitor their ranks for any unprofessional performance, and alert those pilots who exhibit unprofessionalism to its dangers...."[2] The NTSB cited seven previous accidents where a lack of cockpit discipline contributed to a crash. Piedmont Flight 22 was not among those cited.

What began as a recommendation in 1974 morphed into regulation in 1981, when the FAA established the "Sterile Cockpit" rule. This rule states,

    a. No certificate holder shall require, nor may any flight crewmember perform, any duties during a critical phase of flight except those duties required for the safe operation of the aircraft. Duties such as company required calls made for such non-safety related purposes as ordering galley supplies and confirming passenger connections, announcements made to passengers promoting the air carrier or pointing out sights of interest, and filling out company payroll and related records are not required for the safe operation of the aircraft.

b. No flight crew member may engage in, nor may any pilot in command permit, any activity during a critical phase of flight which could distract any flight crew member from the performance of his or her duties or which could interfere in any way with the proper conduct of those duties. Activities such as eating meals, engaging in nonessential conversations within the cockpit and nonessential communications between the cabin and cockpit crews, and reading publications not related to the proper conduct of the flight are not required for the safe operation of the aircraft.

c. For the purposes of this section, critical phases of flight includes all ground operations involving taxi, takeoff and landing, and all other flight operations conducted below 10,000 feet, except cruise flight.[3]

No room for ambiguity existed within the regulation. If a pilot was operating a commercial aircraft beneath ten thousand feet, every aspect of a crew's performance must deal with the safe operation of that aircraft. As Robert Sumwalt, a former vice chairman of the NTSB, wrote, "It's no secret. When a flight crew's attention is diverted from the task of flying, the chance of error increases."[4] Conversely, if a flight crew concentrates on the safe operation of the aircraft, the chance of error decreases. While it took seven years from the tragedy of Flight 212 to the creation of the "Sterile Cockpit" rule, the rule undoubtedly saved many lives and shone a spotlight on cockpit discipline.

But even before the 1974 Flight 212 crash, the NTSB possessed a voice recording of the fatal last seconds of three men bantering about a cigarette fire in their cockpit. Certainly, the actions of the Piedmont crew rose to the same level of seriousness as the banter onboard Flight 212. However, no one at the NTSB used the Flight 22 CVR to initiate a safety program regarding cockpit discipline. They waited seven years to even bring the issue to the public's attention. For seven years, the NTSB possessed this CVR and did not use it to spotlight the problems of cockpit discipline in aircraft. How many lives would have been saved if the NTSB had spotlighted the issues of cockpit discipline seven years before they actually did so?

In 1978, in an accident eerily similar to Flight 22, a Pacific Southwest

Airlines (PSA) Boeing 727 collided with a Cessna 172 above San Diego, California. The death toll of this accident was terrible: 144 persons, including seven on the ground. The probable cause of this accident was the "failure of the flight crew of Flight 182 to comply with the provisions of a maintain-visual-separation clearance...."[5] The ATC controller's actions were also listed as contributing to this accident. But within the report was contained a passage about a dangerous cockpit environment. "Although company procedures urge the flight engineer to plan routine paperwork and radio contact ... to be accomplished at altitudes above 10,000 feet, he was involved with radio contacts with the company when the Cessna was pointed out to Flight 182,"[6] but extraneous conversation did continue down to 3,200 feet.[7] "Although the conversation was not causal, it does point out the dangers inherent in this type of cockpit environment during descent and approach to landing."[8] Certainly, the actions of the Piedmont Flight 22 crew rose to a higher level of inattention than the flight engineer's actions on 182, but again the NTSB in its final report stated there was nothing of a probative value on the Piedmont tape.

Clearly, the NTSB could at the very least have mentioned the actions of the Piedmont crew in its final report. Had it felt they did not rise to the level of a probable cause, other levels of seriousness could have been used without acknowledging liability. The NTSB could have cited the actions of the Piedmont crew as a contributory or causal factor. It could even, as was done in the PSA investigation, have mentioned it in its report and commented on the dangers of lack of cockpit discipline without applying cause. The NTSB did not choose to do so. Rather, it ignored the lessons that could have been taught by the Flight 22 crew's actions and waited seven years, until another tragedy, to give cockpit discipline the level of importance it deserved. It is impossible to determine the number of lives that could have been saved had the NTSB acted sooner and pointed out the unprofessional conduct of the Piedmont Flight 22 crew.

23

"AT THE VERY LEAST, THE CONTROLLER SHOULD HAVE REQUESTED A READ-BACK"

In 2006, the NTSB had an opportunity to renew the investigation of the Flight 22 midair collision. A third-party petition of modification, one of the very few ever granted by NTSB, pointed out the discrepancies in the investigation. After a CBS News report on the original investigation, as well as petitions from concerned local leaders and letters from members of Congress, the NTSB agreed to revisit the original investigation.

The NTSB's image had been tarnished after the TWA Flight 800, Egypt Air 900, and American Airlines Flight 587 investigations between 1996 and 2001. Charges of cover-up and conspiracies dogged their handling of these investigations, in which many people dismissed the board's official findings regarding these accidents. Reopening the Flight 22 case was a rare opportunity for the NTSB to admit, and correct, its mistakes.

In February 2007, the NTSB revealed that it would not overturn the original ruling, agreeing with every aspect of the original investigators' findings. The board validated Thomas Saunders as lead investigator, finding no conflict of interest. It dismissed Captain Schulte's lit cigarette and subsequent lack of cockpit discipline, explaining it away as insignificant. Inexplicably, it argued that Watkins should have relied on Addison's radio transmissions for separation of the aircraft, rather than the other way around.[1] Since no "new" evidence had been uncovered, the NTSB relied on the original investigators' interpretation of the facts, even if those facts were interpreted wrongly.

Even though James Watkins stated, "Three two one Sugar cleared over the VOR to Broad River, correction make that the Asheville radio beacon ... over the VOR to the Asheville radio beacon,"[2] the 2007 decision stated that "the Safety Board agrees with the original investigators and does not find the controller's misstatement to be significant or causal."[3] Watkins did not indicate what heading Addison was to fly to the Asheville radio beacon, nor did he indicate the type of approach in effect at the airport. The NTSB left out of its response that Atlanta control warned Addison that ILS landings were in effect at the airport. Watkins gave no indication to Addison that the approach procedure had changed.

Since Addison acknowledged that he was heading to the Asheville Radio beacon, the NTSB placed all blame upon him. Addison did not know that the Asheville radio beacon and the Broad River radio beacon were not the same. Again, there are over seven locations on the approach plate with the name "Asheville." Without any indication from Watkins as to what direction he was to approach the Asheville radio beacon from, nor any indication that ILS landings were not in effect, the onus could only be shouldered by James Watkins to make sure that the pilots under his control were doing what he wanted them to do.

The board's decision also noted that Watkins had to rely on the Cessna pilot's position reports to maintain separation between aircraft because of lack of radar at the airport. Even Francis McAdams noted the fallacy of this argument. He argued in 1968 that "the pilot under such a system must rely upon the superior knowledge of the controller with respect to other aircraft and the accuracy, adequacy, and timeliness of the clearances.... At the very least, the controller should have requested a read-back from the Cessna."[4] Instead, Watkins never even heard the critical transmission from David Addison in which he stated that he was heading toward the Broad River radio beacon.

The board's reevaluation also "found nothing to indicate the relationship between the investigator in charge and Piedmont's vice president compromised the investigation or the report's content."[5] The Safety Board did not indicate whether or not it felt Thomas Saunders should have been appointed to that specific investigation. Nor did it indicate that this practice has stopped or had not been used subsequent to the Flight 22 investigation. Its excuse for sanctioning such a practice was that "many staff members, not just the investigator in charge, contributed to the report and the Board Members concurred with the findings and probable cause and adopted the report."[6] It released its response in 2007, leaving the original findings intact.

In a September 2006 deposition, Zeke Saunders indicated that once he learned his brother was the investigator-in-charge of the accident investigation, he turned over his responsibilities to other Piedmont employees.[7] While Saunders may have turned over his duties, he still made most of Piedmont's party appointments to the investigation before doing so. As subsequent testimony proved, he was privy to facts regarding the investigation while it was going on. This information was submitted to him by the staff that he appointed to help with the investigation, in violation of NTSB party rules. It also does not matter whether or not Saunders removed himself from the investigation. It would not have made a difference if he took a leave of absence from Piedmont Airlines and traveled around the world during the investigation; he was still the brother of the lead investigator. The only way to eliminate a conflict of interest, or any appearance of a conflict of interest, was for Thomas Saunders to excuse himself from the investigation. Had that happened, it would have eliminated any accusations of a conflict of interest.

Finally, the NTSB addressed the issue of the cigarette fire in the cockpit of the Boeing 727. The board agreed with the findings of the original report that the fire was not significant or causal. In an admission that is as unbelievable as it is appalling, it wrote in the 2007 response that "because the crew laughed about the fire, the original accident investigators did not find the small fire to be significant or causal; after reexamining the case, the Safety Board agrees with that assessment."[8] Thankfully, subsequent NTSB investigators would cite wrongdoing by the crew whether they were laughing or not.

As a point of argument, there was no laughter in the cockpit of Flight 22 during the conversation regarding the fire. The captain was speaking to the copilot when the flight engineer interrupted them by asking them if there was a cigarette fire. Only after Schulte found the source and extinguished it did they make a joke, and the joke was about burnt steaks, not the cigarette fire.

The board also stated that since both pilots of the two aircraft "are responsible for maintaining lookout for other aircraft, it would not be reasonable to single out one flight crew over another on this issue."[9] This is absurd. Only a minute before the accident did James Watkins in the control tower tell David Addison what kind of approach to use for landing, an ADF-2 approach. Subsequent to that command, there can hardly be doubt that Addison was in the process of looking for the ADF-2 approach plate. Being a careful pilot with over ten thousand hours of flight to his credit, he would have reviewed the approach plate and turned his craft toward the north end of the airport. Since the collision took place on the 243 degree dial of the Asheville VOR, there is no doubt that Addison was doing this

23. "Should have requested a read-back" 185

because he left off the Asheville VOR on a 238 degree heading. Turning an aircraft from the southwest to the north would take a pilot directly over the 243 degree radial, where the collision took place. While Addison was performing his duty, the crew of the Piedmont aircraft was distracted by a cigarette fire and cracking jokes about burnt steaks. To even equate the two pilots' actions as being equal could very well be the worst injustice of all the injustices surrounding this entire case.

The board's 2007 decision was another blow to its integrity and reputation as a non-biased, independent government agency. Having ignored the opportunity to state its opposition to conflicts of interest among staff, every investigation it subsequently undertakes will be looked at with a jaundiced eye. Having justified a pilot's lack of discipline in the cockpit because he joked about it tarnishes its reputation as leaders in the field of safety. And wrongly blaming an innocent man (twice) for a tragedy that cost the lives of eighty-two people casts a pall over every accident the NTSB has ever investigated or will investigate. The crash of Flight 22 left a black cloud lingering in the sky for several hours after the accident. The investigation of the accident has left a black cloud lingering over the NTSB for forty years.

Chapter Notes

Chapter 1

1. National Transportation Safety Board, *Complete Transcript of the Last 18 Minutes of the Cockpit Voice Recorder Tape Recording Aboard Piedmont Boeing 727 Aircraft N68650*, July 19, 1967, 3.
2. *Ibid.*
3. "An Aircraft Accident Involving Piedmont Airlines, Inc., Boeing 727, N68650 and Lanseair Cessna 310, N3121S at Hendersonville, North Carolina, July 19, 1967," public hearing (Washington, D.C.: Department of Transportation, Docket No. SA-400, 1967), 627.
4. *Ibid.*
5. John D. Rawson, *Group Chairman's Factual Report of Investigation*, National Transportation Safety Board, SA-400, Exhibit No. 11A, 1967, 2–3.
6. Aircraft: Boeing 727–100, http://aeroweb.org/specs/boeing/727-100 (accessed September 9, 2012).
7. *Winston-Salem Journal*, July 20, 1967.
8. *Ibid.*
9. Edwin Nelmes, *Operations Group Chairman's Factual Report*, National Transportation Safety Board, SA-400, Exhibit No. 2A, 1967, 13–14.
10. *Ibid.*, 14.
11. *Ibid.*
12. Aircraft Accident Report, "Piedmont Aviation, Inc. Piedmont Airlines Division Boeing 727, N68650, Lanseair Inc., Cessna 310, N3121S Mid Air Collision, Hendersonville, North Carolina, July 19, 1967" (Washington, D.C.: National Transportation Safety Board, SA-400, File 1–0005, 1968), Appendix A.
13. Gerrit Walhout, *Human Factors Group Chairman's Factual Report*, National Transportation Safety Board, SA-400, Exhibit No. 6A, 1967, 7.
14. Edwin Nelmes, *Operations Group Chairman's Factual Report*, National Transportation Safety Board, SA-400, Exhibit No. 2A, 1967, 14.
15. Gerrit Walhout, *Human Factors Group Chairman's Factual Report*, National Transportation Safety Board, SA-400, Exhibit No. 6A, 1967, 8.
16. Edwin Nelmes, *Operations Group Chairman's Factual Report*, National Transportation Safety Board, SA-400, Exhibit No. 2A, 1967, 15.
17. Gerrit Walhout, *Human Factors Group Chairman's Factual Report*, National Transportation Safety Board, SA-400, Exhibit No. 6A, 1967, 8.
18. Edwin Nelmes, *Operations Group Chairman's Factual Report*, National Transportation Safety Board, SA-400, Exhibit No. 2A, 1967, 15.
19. National Transportation Safety Board, *Complete Transcript of the Last 18 Minutes of the Cockpit Voice Recorder Tape Recording Aboard Piedmont Boeing 727 Aircraft N68650*, July 19, 1967, 5.
20. *Ibid.*
21. *Ibid.*
22. *Ibid.*
23. Edwin Nelmes, *Operations Group Chairman's Factual Report*, National Transportation Safety Board, SA-400, Exhibit No. 2A, 1967, 16.
24. National Transportation Safety Board, *Complete Transcript of the Last 18 Minutes of the Cockpit Voice Recorder Tape Recording Aboard Piedmont Boeing 727 Aircraft N68650*, July 19, 1967, 5. Holly Anderson Case, interview by author July 17, 2004, Hendersonville, North Carolina.
25. Gerrit Walhout, *Human Factors Group Chairman's Factual Report*, National Transportation Safety Board, SA-400, Exhibit No. 6A, 1967, 9.
26. Andrew J. Davis, Untitled, To Whom It May Concern, Weather Conditions in South-

eastern United States, July 20, 1967, photocopied.
27. Benjamin M. Folk, SATCS, "Narrative Statement" (Charlotte, NC: Department of Transportation, July 19, 1967), photocopied.
28. Roys C. Jones, "Analysis of Air Traffic Control Exercised on N3121S, A Cessna 310 Aircraft and Piedmont Flight 22, A Boeing 727 Aircraft on July 19, 1967, Prior to Their Mid-Air Collision Near Hendersonville, North Carolina at 1600 Hours, One Minute and Sixteen Seconds, GMT" (August 10, 1967, photocopied), 2.
29. Ronald A. Painter, ATCS, "Narrative Statement" (Charlotte, NC: Department of Transportation, July 19, 1967), photocopied.
30. Department of Transportation, *Transcript of Recorder Number One Channel Number Two Ground Control Position*, Airport Traffic Control Tower, Charlotte, North Carolina, July 19, 1967, 2.
31. *Ibid.*
32. *Ibid.*, 3.
33. *FAA, Air Ground Communications Transcript, Atlanta Center*. (Washington, D.C.: National Transportation Safety Board, SA-400), 2.
34. *Ibid.*
35. *Ibid.*
36. *Ibid.*
37. Graves, 4.
38. *Ibid.*

Chapter 2

1. Sally Taylor, interview by author June 30, 2007, Hendersonville, North Carolina.
2. Lance Benham, interview by author May 20, 2006, telephone.
3. Mrs. Shuler, interviewed by author July 17, 2004, Hendersonville, NC.
4. Benjamin T. Harrison and Christopher Mosher, "John T. McNaughton and Vietnam: The Early Years as Assistant Secretary of Defense, 1964–1965," *History* 9 (October 2007): 503.
5. Robert McNamara and Brian VanDeMark, *In Retrospect* (New York: Times Books, 1995), 280.
6. *Ibid.*

Chapter 3

1. J.A.C. Dunn, "The History of Piedmont Airlines," *Pace* (December 1988): 59.
2. "Aircraft Accident Report," Civil Aeronautics Board, File No. 1–0065, SA-348, 1961, 12.
3. "One Man's Anguish," *Time* (May 5, 1961): 17.
4. *Ibid.*
5. "Aircraft Accident Report," Civil Aeronautics Board, File No. 1–0065, SA-348, 1961, 11.
6. *Ibid.*, 12.
7. J.A.C. Dunn, "The History of Piedmont Airlines," *Pace* (December 1988): 66.
8. *Ibid.*, 67.
9. Richard Hurley, "The Passing of the Pacemaker," *Airliners* (Fall 1988): 43.
10. *Ibid.*, 44.
11. J.A.C. Dunn, "The History of Piedmont Airlines," *Pace* (December 1988): 69.
12. Francis X. Graves, *ATC Group Chairman's Factual Report*, National Transportation Safety Board, SA-400, Exhibit No. 3A, 1967, 14.
13. Federal Aviation Administration, *Transcription of Recorder Channel Number Two Approach Control/Flight Data Position*, Airport Traffic Control Tower: Asheville, North Carolina, July 19, 1967, 14–15.
14. *Ibid.*, 15.
15. National Transportation Safety Board, *Complete Transcript of the Last 18 Minutes of the Cockpit Voice Recorder Tape Recording Aboard Piedmont Boeing 727 Aircraft N68650*, July 19, 1967, 8.
16. Roys C. Jones, "Analysis of Air Traffic Control Exercised on N3121S, A Cessna 310 Aircraft and Piedmont Flight 22, A Boeing 727 Aircraft on July 19, 1967, Prior to Their Mid-Air Collision Near Hendersonville, North Carolina at 1600 Hours, One Minute and Sixteen Seconds, GMT" (August 10, 1967, photocopied), 5.
17. *Ibid.*
18. Federal Aviation Administration, *Transcription of Recorder Channel Number Two Approach Control/Flight Data Position*, Airport Traffic Control Tower: Asheville, North Carolina, July 19, 1967, 15.
19. National Transportation Safety Board, *Complete Transcript of the Last 18 Minutes of the Cockpit Voice Recorder Tape Recording Aboard Piedmont Boeing 727 Aircraft N68650*, July 19, 1967, 9.
20. *Ibid.*
21. *Ibid.*, 10.
22. *Ibid.*
23. *Ibid.*, 11.
24. *Ibid.*
25. Federal Aviation Administration, *Transcription of Recorder Channel Number Two Approach Control/Flight Data Position*, Airport Traffic Control Tower: Asheville, North Carolina, July 19, 1967, 17.
26. *Ibid.*

27. Hearings Before the Committee on Commerce, United States Senate, Eighty-ninth Congress, First Session, on Nomination of Gen. William F. McKee, Administrator, Federal Aviation Agency, Warren G. Magnuson, chairman (Washington, D.C.: Government Printing Office, 1965), 50–55.
28. *New York Times*, February 6, 1966.
29. *Ibid*.
30. Lyndon B. Johnson, The President's News Conference at Austin, TX, December 6, 1966, accessed May 11, 2011, from www.presidency.ucsb.edu.
31. R.H. Watts, *Accident Reports: Development and Use, Collection, Recording, Retrieval and Dissemination* (Montreal: International Civil Aviation Organization, undated), Session 7.
32. Stanley Lyman, House of Representatives, Subcommittee on Transportation and Aeronautics, Committee on Interstate and Foreign Commerce, Samuel N. Friedel, chairman (Washington, D.C.: Government Printing Office, 1968), 283.
33. *New York Times*, May 13, 1967.
34. *New York Times*, April 30, 1967.
35. *New York Times*, May 13, 1967.
36. *Ibid*.
37. *Ibid*.
38. *Ibid*.

Chapter 4

1. Aircraft Accident Report, "Piedmont Aviation, Inc. Piedmont Airlines Division Boeing 727, N68650, Lanseair Inc., Cessna 310, N3121S Mid Air Collision, Hendersonville, North Carolina, July 19, 1967" (Washington, D.C.: National Transportation Safety Board, SA-400, File 1–0005, 1968), 9–10.
2. *Ibid.*, 24.
3. National Transportation Safety Board, *Complete Transcript of the Last 18 Minutes of the Cockpit Voice Recorder Tape Recording Aboard Piedmont Boeing 727 Aircraft N68650*, July 19, 1967, 11.
4. *Ibid*.
5. *Ibid*.
6. *Ibid*.
7. *Ibid*.
8. *Ibid*.
9. *Ibid.*, 12.
10. *Ibid*.
11. *Ibid*.
12. *Ibid.*, 12–13.
13. *Ibid.*, 13.
14. *Ibid*.
15. *Ibid*.
16. *Ibid*.
17. *Ibid*.
18. Aircraft Accident Report, "United Airlines, Inc., B-727, N7036U, in Lake Michigan" (Washington, D.C.: National Transportation Safety Board, File–1–0030, 1967), 37.
19. Aircraft Accident Report, "American Airlines, Inc., Boeing 727, N1996, Near the Greater Cincinnati Airport, Constance, Kentucky, November 8, 1965" (Washington, D.C.: Civil Aeronautics Board, SA-387, File No. 1–0031, 1966), 26.
20. Aircraft Accident Report, "United Airlines, Inc., Boeing 727, N7030U, Salt Lake City, Utah, November 11, 1965" (Washington, D.C.: Civil Aeronautics Board, SA-388, File No. 1–0032), 14.
21. *New York Times*, November 14, 1965.
22. *Ibid*.
23. *Ibid*.
24. *New York Times*, February 16, 1966.
25. *New York Times*, February 18, 1966.
26. *Ibid*.
27. Aircraft Accident Report, "American Airlines, Inc., Boeing 727, N1996, Near the Greater Cincinnati Airport, Constance, Kentucky, November 8, 1965" (Washington, D.C.: Civil Aeronautics Board, SA-387, File No. 1–0031, 1966), 25.
28. *Ibid*.
29. *Ibid.*, 26.
30. Aircraft Accident Report, "Braniff Airlines, Inc., Lockheed Electra, N9705C, Buffalo, Texas, September 29, 1959" (Washington, D.C.: Civil Aeronautics Board, 1961), 1.
31. Aircraft Accident Report, "Northwest Airlines, Lockheed Electra, N121US, Near Cannelton, Indiana, March 17, 1960" (Washington, D.C.: Civil Aeronautics Board, 1961), 1.
32. Paul O'Neil, "Brilliant Detection in Jet Age Mystery," *Life*, 25 July 1960, 82, 84.
33. *New York Times*, February 10, 1993.
34. Paul O'Neil, "Brilliant Detection in Jet Age Mystery," *Life*, 25 July 1960, 84.
35. *Ibid*.
36. *Ibid*.
37. *Ibid.*, 85.
38. *Ibid.*, 87.
39. *Ibid*.
40. *Ibid.*, 88.
41. *Ibid.*, 87.
42. *Winston-Salem Journal*, July 20, 1967.
43. *Ibid*.
44. Aircraft Accident Report, "Northwest Airlines, Inc., Boeing 720B, N724US, Near Miami, Florida, February 12, 1963" (Washington, D.C.: Civil Aeronautics Board, SA-372, File No. 1–0006, 1965), 24.
45. Aircraft Accident Report, "Pan American World Airways, Inc., Boeing 707–121,

N709-PA, Near Elkton, Maryland, December 8, 1963" (Washington, D.C.: Civil Aeronautics Board, SA-376, File No.1–0015), 13.
  46. Aircraft Accident Report, "Eastern Airlines Inc., Douglas DC-8, N8607, New Orleans, Louisiana, February 25, 1964" (Washington, D.C.: Civil Aeronautics Board, SA-379, File No. 1–0006, 1966), 1.
  47. Warren R. Young, "Turbulence-Hidden Giant in the Sky," *Life*, 18 December 1964, 92.
  48. Federal Aviation Agency, Federal Aviation Regulations, 14 CFR, Section 121.357, December 31, 1964.
  49. *Ibid.*
  50. John D. Rawson, *Group Chairman's Factual Report of Investigation*, National Transportation Safety Board, SA-400, Exhibit No. 11A, 1967, 10.
  51. *Ibid.*, 13.
  52. *Ibid.*, 14.
  53. *Ibid.*, 16.
  54. *Ibid.*, 20.
  55. House of Representatives, Committee on Interstate and Foreign Commerce, Harley O. Staggers, Chairman (Washington, D.C.: Government Printing Office, 1968), 202.
  56. John D. Rawson, *Group Chairman's Factual Report of Investigation*, National Transportation Safety Board, SA-400, Exhibit No. 11A, 1967, 23.
  57. "An Aircraft Accident Involving Piedmont Airlines, Inc., Boeing 727, N68650 and Lanseair Cessna 310, N3121S at Hendersonville, North Carolina, July 19, 1967," public hearing (Washington, D.C.: Department of Transportation, Docket No. SA-400, 1967), 638.
  58. *Ibid.*, 638–639.

Chapter 5

  1. Federal Aviation Administration, *Transcription of Recorder Channel Number Two Approach Control/Flight Data Position*, Airport Traffic Control Tower, Asheville, North Carolina, July 19, 1967, 17.
  2. *Ibid.* The transcription of the garbled transmission has Addison saying "Oh good shit." This may have been the closest approximation of an unintelligible phrase. It may also have been exactly what Addison said in a muttered aside to his passenger as they searched the charts in vain for the non-existent radio beacon.
  3. *Ibid.*
  4. *Ibid.*
  5. *Ibid.*
  6. National Transportation Safety Board, *Complete Transcript of the Last 18 Minutes of the Cockpit Voice Recorder Tape Recording Aboard Piedmont Boeing 727 Aircraft N68650*, July 19, 1967, 14.
  7. *Ibid.*
  8. *Ibid.*
  9. *Ibid.*
  10. *Ibid.*, 15.
  11. Federal Aviation Administration, *Transcription of Recorder Channel Number One Local/Ground Control Position*, Airport Traffic Control Tower, Asheville, North Carolina, July 19, 1967, 8.
  12. Federal Aviation Agency, Federal Aviation Regulations, 14 CFR, Section 91.87 (in effect in 1967).
  13. National Transportation Safety Board, *Complete Transcript of the Last 18 Minutes of the Cockpit Voice Recorder Tape Recording Aboard Piedmont Boeing 727 Aircraft N68650*, July 19, 1967, 16.
  14. *Ibid.*
  15. Federal Aviation Administration, *Transcription of Recorder Channel Number Two Approach Control/Flight Data Position*, Airport Traffic Control Tower, Asheville, North Carolina, July 19, 1967, 17–18.
  16. *Ibid.*, 18.
  17. National Transportation Safety Board, *Complete Transcript of the Last 18 Minutes of the Cockpit Voice Recorder Tape Recording Aboard Piedmont Boeing 727 Aircraft N68650*, July 19, 1967, 17.
  18. *Ibid.*
  19. *Ibid.*
  20. *Ibid.*, 18.
  21. *Ibid.*

Chapter 6

  1. Bob Terrell, *Historic Asheville (1792–1930)* (North Carolina: WorldComm, 1997), 16.
  2. Nan Chase, *Asheville: A History* (Jefferson, NC: McFarland, 2007), 9.
  3. National Park Service, "National Register of Historic Places Travel Itinerary," http://www.nps.gov/nr/travel/asheville/architecture.htm, accessed August 21, 2013.
  4. Richard Starnes, *Tourism and Society in Western North Carolina: Creating the Land of the Sky* (Tuscaloosa: University of Alabama Press, 2005), 108.
  5. *Times-News* (North Carolina), January 16, 1961.
  6. Lacy Griffin, personal conversation, 2002.
  7. Mr. Harry LaViers, Interview by the National Transportation Safety Board, "Wit-

ness Statement of Mid-Air Collision," Hendersonville, North Carolina, July 24, 1967.
8. *Ibid.*
9. *Ibid.*
10. Verne R. Davis, Interview by the National Transportation Safety Board, "Witness Statement of Mid-Air Collision," Hendersonville, North Carolina, July 23, 1967. *Ibid.*
11. *Ibid.*
12. *Ibid.*
13. *Ibid.*
14. Joel Dermid, Interview by the National Transportation Safety Board, "Witness Statement of Mid-Air Collision," Hendersonville, North Carolina, July 21, 1967. *Ibid.*
15. *Ibid.*
16. Mrs. Robert C. Wilson, Interview by the National Transportation Safety Board, "Witness Statement of Mid-Air Collision," Hendersonville, North Carolina, July 20, 1967.
17. *Ibid.*
18. *Ibid.*
19. Roy Beddingfield, Interview by the National Transportation Safety Board, "Witness Statement of Mid-Air Collision," Hendersonville, North Carolina, July 21, 1967.
20. *Ibid.*
21. William Murray, Interview by the National Transportation Safety Board, "Witness Statement of Mid-Air Collision," Hendersonville, North Carolina, July 21, 1967.
22. Robert Kaufman, interview by author April 10, 2006, Miami, Florida.
23. *Atlanta Constitution*, July 20, 1967.

Chapter 7

1. Gerrit Walhout, *Human Factors Group Chairman's Factual Report*, National Transportation Safety Board, SA-400, Exhibit No. 6A, 1967, 3.
2. *Atlanta Journal-Constitution*, July 20, 1967.
3. Robert Kaufman, interview by author April 10, 2006, Miami, Florida.
4. James T. Childs, *Structures Group Chairman's Factual Report*, National Transportation Safety Board, SA-400, Exhibit No. 7A, 1967, 1–3.
5. *Ibid.*
6. J.J. Barnes, Interview by the National Transportation Safety Board, "Witness Statement of Mid-Air Collision," Hendersonville, North Carolina, August 7, 1967.
7. Federal Aviation Administration, *Transcription of Recorder Channel Number One Local/Ground Control Position*, Airport Traffic Control Tower, Asheville, North Carolina, July 19, 1967, 9.

8. *Ibid.*
9. *Ibid.*
10. Federal Aviation Administration, *Transcription of Recorder Channel Number Two Approach Control/Flight Data Position*, Airport Traffic Control Tower, Asheville, North Carolina, July 19, 1967, 19.
11. Federal Aviation Administration, *Transcription of Recorder Channel Number One Local/Ground Control Position*, Airport Traffic Control Tower, Asheville, North Carolina, July 19, 1967, 9.
12. Federal Aviation Administration, *Transcription of Recorder Channel Number Two Approach Control/Flight Data Position*, Airport Traffic Control Tower, Asheville, North Carolina, July 19, 1967, 19.
13. Federal Aviation Administration, *Transcription of Recorder Channel Number One Local/Ground Control Position*, Airport Traffic Control Tower, Asheville, North Carolina, July 19, 1967, 9.
14. *Ibid.*
15. *Ibid.*
16. *Ibid.*
17. *Ibid.*
18. *Ibid.*
19. *Ibid.*
20. *Ibid.*, 10.
21. Federal Aviation Administration, *Transcription of Recorder Channel Number Two Approach Control/Flight Data Position*, Airport Traffic Control Tower, Asheville, North Carolina, July 19, 1967, 19.
22. *Ibid.*
23. *Ibid.*
24. *Ibid.*
25. Federal Aviation Administration, *Transcription of Recorder Channel Number One Local/Ground Control Position*, Airport Traffic Control Tower, Asheville, North Carolina, July 19, 1967, 10.
26. *Ibid.*
27. *Ibid.*
28. *Ibid.*
29. N.H. Van Sicklin, Interview by the National Transportation Safety Board, "Witness Statement of Mid-Air Collision," Hendersonville, North Carolina, July 21, 1967.
30. Federal Aviation Administration, *Transcription of Recorder Channel Number One Local/Ground Control Position*, Airport Traffic Control Tower, Asheville, North Carolina, July 19, 1967, 10–11.
31. Federal Aviation Administration, *Transcription of Recorder Channel Number Two Approach Control/Flight Data Position*, Airport Traffic Control Tower, Asheville, North Carolina, July 19, 1967, 19.
32. *Ibid.*

33. Federal Aviation Administration, *Transcription of Recorder Channel Number One Local/ Ground Control Position*, Airport Traffic Control Tower, Asheville, North Carolina, July 19, 1967, 11.
34. *Ibid.*
35. *Ibid.*
36. *Ibid.*
37. *Ibid.*
38. *Ibid.*
39. Federal Aviation Administration, *Transcription of Recorder Channel Number Two Approach Control/Flight Data Position*, Airport Traffic Control Tower, Asheville, North Carolina, July 19, 1967, 20.
40. *Ibid.*
41. *Transcript of Conversations Recorded at the Federal Aviation Administration Air Traffic Control Tower, Asheville Municipal Airport, Arden, North Carolina*, July 19, 1967.
42. *Ibid.*
43. *Ibid.*
44. *Ibid.*
45. *Ibid.*
46. *Ibid.*
47. *Ibid.*
48. *Ibid.*
49. *Ibid.*
50. *Ibid.*
51. *Ibid.*
52. *Ibid.*
53. *Ibid.*
54. *Ibid.*
55. *Ibid.*
56. *Ibid.*
57. *Ibid.*
58. *Ibid.*
59. *Ibid.*
60. *Ibid.*
61. *Ibid.*
62. *Ibid.*
63. *Ibid.*
64. *Ibid.*

Chapter 8

1. Whitney Gillilland, Speech Given at the University of Miami (Iowa City: University of Iowa, 1968).
2. Hearings Before the Committee on Commerce, United States Senate, Ninetieth Congress, First Session, on Nominations of Joseph P. O'Connell, Francis H. McAdams, Rear Adm. Louis N. Thayer, John H. Reed, Oscar Laurel, Warren G. Magnuson, Chairman (Washington, D.C.: Government Printing Office, 1967), 1.
3. *Ibid.*
4. *Ibid.*, 10.
5. *Ibid.*
6. *Ibid.*, 3.
7. *Ibid.*, 4.
8. *Ibid.*, 5.
9. *Ibid.*, 12.
10. *Ibid.*, 16.
11. Victor Kayne, *House of Representatives, Subcommittee on Transportation and Aeronautics, Committee on Interstate and Foreign Commerce*, Samuel N. Friedel, Chairman (Washington, D.C.: Government Printing Office, 1968), 188.
12. Joseph J. O'Connell, *House of Representatives, Committee on Interstate and Foreign Commerce*, Harley O. Staggers, Chairman (Washington, D.C.: Government Printing Office, 1968), 5.
13. John R. Reagan, Deposition, Western District of North Carolina, March 12, 1969, 2699.
14. Frank C. Nicholson, Deposition, Western District of North Carolina, April 29, 1969, 4522.
15. Harold Kimball Saunders, Sr., Affidavit in Support of US Airways [Successor in interest to Piedmont Airlines], Response to Petition of Modification filed March 2005 by Paul D. Houle, Forsyth County, North Carolina, September 5, 2006.
16. Francis X. Graves, *ATC Group Chairman's Factual Report*, National Transportation Safety Board, SA-400, Exhibit No. 3A, 1967, 1.
17. Rudolf Kapustin, *PowerPlant Group Chairman's Factual Report*, National Transportation Safety Board, SA-400, Exhibit No. 8A, 1967, 1.
18. Hubert McCaleb, *Weather Group Chairman's Factual Report*, National Transportation Safety Board, SA-400, Exhibit No. 5A, 1967, 1.
19. William Weston, *Group Chairman's Factual Report*, National Transportation Safety Board, SA-400, Exhibit No. 9A, 1967, 1.
20. O.E. Patton, *Flight Recorder Chairman's Factual Report*, National Transportation Safety Board, SA-400, Exhibit No. 10A, 1967, 1.
21. John D. Rawson, *Group Chairman's Factual Report*, National Transportation Safety Board, SA-400, Exhibit No. 11A, 1967, 1.
22. Edwin Nelmes, *Operations Group Chairman's Factual Report*, National Transportation Safety Board, SA-400, Exhibit No. 2A, 1967, 1.
23. Holly Anderson Case, interview by author July 17, 2004, Hendersonville, North Carolina.
24. *Times-News* (North Carolina), July 21, 1967.
25. *Springfield Leader-Press* (Missouri), July 20, 1967.

26. *Ibid.*
27. Lyndon B. Johnson, "Statement by the President on the Death of John T. McNaughton," July 19, 1967. Online by Gerhard Peters and John T. Woolley, *The American Presidency Project*, http://www.presidency.ucsb.edu/ws/?pid=28357, accessed May 11, 2011.
28. Robert McNamara: "Memo to the President," May 19, 1967, accessed July 6, 2013, from www.lbjlibrarytumblr.com.
29. Stanley Lyman, *House of Representatives, Subcommittee on Transportation and Aeronautics, Committee on Interstate and Foreign Commerce*, Samuel N. Friedel, Chairman (Washington, D.C.: Government Printing Office, 1968), 287.
30. Hearings Before the Committee on Commerce, United States Senate, Ninetieth Congress, First Session, on Nominations of Joseph P. O'Connell, Francis H. McAdams, Rear Adm. Louis N. Thayer, John H. Reed, Oscar Laurel, Warren G. Magnuson, Chairman (Washington, D.C.: Government Printing Office, 1967), 2.
31. Stanley Lyman, *House of Representatives, Subcommittee on Transportation and Aeronautics, Committee on Interstate and Foreign Commerce*, Samuel N. Friedel, Chairman (Washington, D.C.: Government Printing Office, 1968), 283.
32. *Springfield Leader-Press* (Missouri) July 24, 1967.
33. *Ibid.*
34. *Ibid.*
35. *Ibid.*
36. Drew Pearson, "Washington Merry-Go-Round" (Bell-McClure Syndicate), July 6, 1967.
37. Hearings Before the Committee on Commerce, United States Senate, Ninetieth Congress, First Session, on Nominations of Joseph P. O'Connell, Francis H. McAdams, Rear Adm. Louis N. Thayer, John H. Reed, Oscar Laurel, Warren G. Magnuson, Chairman (Washington, D.C.: Government Printing Office, 1967), 2.
38. *New York Times*, July 21, 1967.

Chapter 9

1. *Winston-Salem Journal*, July 20, 1967.
2. *House of Representatives, Committee on Interstate and Foreign Commerce*, Harley Staggers, Chairman (Washington, D.C.: Government Printing Office, 1968), 3.
3. *Ibid.*, 5.
4. *Ibid.*
5. *Ibid.*
6. *Ibid.*, 6.
7. *Ibid.*
8. *Ibid.*
9. *Ibid.*
10. Bobbie Allen, *House of Representatives, Committee on Interstate and Foreign Commerce*, Harley Staggers, Chairman (Washington, D.C.: Government Printing Office, 1968), 7.
11. *Ibid.*
12. *Ibid.*
13. *Ibid.*, 8.
14. *Ibid.*
15. *Ibid.*
16. *Ibid.*, 9.
17. *Ibid.*
18. *Ibid.*
19. *Ibid.*
20. *Ibid.*
21. *Ibid.*, 10.
22. *Ibid.*
23. *Ibid.*, 11.
24. *Ibid.*

Chapter 10

1. William McKee, *House of Representatives, Committee on Interstate and Foreign Commerce*, Harley Staggers, Chairman (Washington, D.C.: Government Printing Office, 1968), 18.
2. *Ibid.*, 19.
3. *Ibid.*
4. *Ibid.*
5. *Ibid.*, 19–20.
6. *Ibid.*, 20.
7. *Ibid.*
8. *Ibid.*, 21.
9. *Ibid.*
10. *Ibid.*
11. *Ibid.*
12. *Ibid.*
13. *Ibid.*
14. *Ibid.*, 22.
15. William McKee, *House of Representatives, Committee on Interstate and Foreign Commerce*, Harley Staggers, Chairman (Washington, D.C.: Government Printing Office, 1968), 22.
16. *Ibid.*
17. *Ibid.*
18. *Ibid.*, 23.
19. *Ibid.*
20. *Ibid.*
21. *Ibid.*
22. *Ibid.*
23. *Ibid.*, 24.
24. *Ibid.*
25. *Ibid.*
26. *Ibid.*
27. *Ibid.*
28. *Ibid.*
29. *Ibid.*, 30.
30. *Ibid.*
31. *Ibid.*

32. *Ibid.*
33. *Ibid.*
34. *Ibid.*
35. *Ibid.*
36. *Ibid.*
37. *Ibid.*
38. *Ibid.*
39. *Ibid.*
40. *Ibid.*
41. *Ibid.*
42. *Ibid.*
43. *Ibid.*
44. *Ibid.*
45. *Ibid.*
46. John F. Kennedy, "Remarks at Colorado Springs to the Graduating Class of the U.S. Air Force Academy," June 5, 1963, accessed May 11, 2011, from www.presidency.ucsb.edu.
47. Don Bedwell, "Going Nowhere Fast," *American Heritage*, Winter 2006, Volume 21, Issue 3.
48. Federal Aviation Agency, *Supersonic Transport Development Program*, 1967, 1.
49. *Ibid.*, 13.
50. *Ibid.*, 3.
51. *Ibid.*, 2.
52. Paul N. Borsky, "Community Reactions to Sonic booms, Oklahoma City Area, February–July 1964, Part 1 Major Findings," National Opinion Research Center, University of Chicago, January 1965, 7.
53. Don Dwiggins, *The SST, Here it Comes, Ready or Not* (New York: Doubleday, 1968), 72–73.
54. Paul N. Borsky, "Community Reactions to Sonic booms, Oklahoma City Area, February–July 1964, Part 1 Major Findings," National Opinion Research Center, University of Chicago, January 1965, 2.
55. *Ibid.*, 2.
56. *The Oklahoman*, February 20, 1966.
57. Lyndon B. Johnson, "Remarks at the Swearing In of General McKee as Administrator, Federal Aviation Agency, July 1, 1965," accessed May 11, 2011, from www.presidency.ucsb.edu.
58. *Ibid.*
59. "SST Price & Progress," *Time*, 30 September 1966, 110.
60. Don Dwiggins, *The SST: Here it Comes, Ready or Not* (New York: Doubleday, 1968), 171.
61. *Ibid.*, 170–171.
62. William McKee, *House of Representatives, Committee on Interstate and Foreign Commerce*, Harley Staggers, Chairman (Washington, D.C.: Government Printing Office, 1968), 31.
63. *Ibid.*
64. *Ibid.*
65. *Ibid.*
66. *Ibid.*
67. *Ibid.*
68. *Ibid.*
69. *Ibid.*
70. *Ibid.*
71. *Ibid.*
72. *Ibid.*
73. *Ibid.*, 35.
74. *Ibid.*
75. *Ibid.*
76. *Ibid.*, 36.
77. *Ibid.*
78. *Ibid.*
79. *Ibid.*
80. *Ibid.*
81. *Ibid.*
82. *Ibid.*, 38.
83. *Ibid.*
84. *Ibid.*
85. *Ibid.*
86. *Ibid.*
87. *Ibid.*
88. *Ibid.*, 39.
89. *Ibid.*
90. *Ibid.*
91. *Ibid.*
92. *Ibid.*
93. *Ibid.*
94. *Ibid.*
95. *Ibid.*
96. *Ibid.*
97. *Ibid.*
98. *Ibid.*
99. *Ibid.*
100. *Ibid.*

Chapter 11

1. Benjam T. Harrison and Christopher L. Mosher, "John T. McNaughton and Vietnam: The Early Years as Assistant Secretary of Defense, 1964–1965," *History, The Journal of the Historical Association* 92 (October 2007): 496–514.
2. *Ibid.*
3. Office of the Secretary of Defense, Vietnam Task Force, "United States-Vietnam Relations: 1945–1967," Volume 1, The Air War in Vietnam, 28.
4. Benjam T. Harrison and Christopher L. Mosher, "John T. McNaughton and Vietnam: The Early Years as Assistant Secretary of Defense, 1964–1965," *History, The Journal of the Historical Association* 92 (October 2007): 507.
5. Office of the Secretary of Defense, Vietnam Task Force, "United States-Vietnam Relations: 1945–1967," Volume 1, The Air War in Vietnam, 28.
6. Lyndon B. Johnson, "The President's

News Conference at the LBJ Ranch," December 31, 1966, accessed May 11, 2011, from www.presidency.ucsb.edu.
 7. Lyndon B. Johnson, "The President's News Conference at Austin, TX, December 6, 1966," accessed May 11, 2011, from www.presidency.ucsb.edu.
 8. George C. Herring, ed., *The Pentagon Papers* (New York: McGraw-Hill, 1993), 174.
 9. Benjam T. Harrison and Christopher L. Mosher, "John T. McNaughton and Vietnam: The Early Years as Assistant Secretary of Defense, 1964–1965," *History, The Journal of the Historical Association* 92 (October 2007): 513.
 10. *Ibid.*, 500.
 11. *Aircraft Accident Report*, "United Air Lines, Inc, DC-8, N 8013U, and Trans World Airlines, Inc., Constellation 1049A, N 6907C, Near Staten Island, New York, December 16, 1960" (Washington, D.C.: Civil Aeronautics Board, SA-361; File 1–0083, 1962), 8.
 12. *Ibid.*, 23.
 13. Richard Lawrence Ottinger, *Biographical Directory of the United States Congress, 1774–Present*, 1929-.
 14. Richard L. Ottinger, *House of Representatives, Committee on Interstate and Foreign Commerce*, Harley Staggers, Chairman (Washington, D.C.: Government Printing Office, 1968), 74.
 15. *Ibid.*
 16. *Ibid.*
 17. William McKee, *House of Representatives, Committee on Interstate and Foreign Commerce*, Harley Staggers, Chairman (Washington, D.C.: Government Printing Office, 1968), 74.
 18. Richard L. Ottinger, *House of Representatives, Committee on Interstate and Foreign Commerce*, Harley Staggers, Chairman (Washington, D.C.: Government Printing Office, 1968), 74.
 19. *Ibid.*
 20. *Ibid.*
 21. *Ibid.*
 22. *Ibid.*
 23. *Ibid.*
 24. *Ibid.*, 74–75.
 25. *Ibid.*, 75.
 26. *Ibid.*, 75.
 27. *Ibid.*, 76.
 28. William McKee, *House of Representatives, Committee on Interstate and Foreign Commerce*, Harley Staggers, Chairman (Washington, D.C.: Government Printing Office, 1968), 76.
 29. *Ibid.*
 30. *Ibid.*
 31. *Ibid.*
 32. Richard L. Ottinger, *House of Representatives, Committee on Interstate and Foreign Commerce*, Harley Staggers, Chairman (Washington, D.C.: Government Printing Office, 1968), 76.
 33. *Ibid.*
 34. *Ibid.*
 35. William McKee, *House of Representatives, Committee on Interstate and Foreign Commerce*, Harley Staggers, Chairman (Washington, D.C.: Government Printing Office, 1968), 76.
 36. Harley Staggers, *House of Representatives, Committee on Interstate and Foreign Commerce*, Harley Staggers, Chairman (Washington, D.C.: Government Printing Office, 1968), 76.
 37. *Ibid.*
 38. William McKee, *House of Representatives, Committee on Interstate and Foreign Commerce*, Harley Staggers, Chairman (Washington, D.C.: Government Printing Office, 1968), 76.

Chapter 12

 1. Stephen Barlay, *Air Crash Detective: An International Report on the Quest for Air Safety* (London: Hamish Hamilton, 1969), 58.
 2. John Robert Reagan, Deposition Held at the Federal Courthouse, Western District of North Carolina, March 15, 1969, 2885.
 3. *Ibid.*
 4. *Ibid.*, 2852.
 5. O.E. Patton, *Flight Recorder Group Chairman's Factual Report*, National Transportation Safety Board, SA-400, Exhibit No. 10A, 1967, 1.
 6. *Ibid.*, 2.
 7. *Ibid.*, 3.
 8. Federal Aviation Agency, *Standard Instrument Approach Procedure-Type ILS*, Asheville Municipal Airport, 16 November 1963.
 9. John D. Rawson, *Group Chairman's Factual Report of Investigation*, National Transportation Safety Board, SA-400, Exhibit No. 11A, 1967, 1.
 10. *Ibid.*
 11. Francis X. Graves, *ATC Group Chairman's Factual Report*, National Transportation Safety Board, SA-400, Exhibit No. 3A, 1967), and Francis X. Graves, *Cockpit Voice Recorder Specialist's Factual Report*, National Transportation Safety Board, SA-400, Exhibit No. 12A, 1967.
 12. Francis X. Graves, *ATC Group Chairman's Factual Report*, National Transportation Safety Board, SA-400, Exhibit No. 3A, 1967.
 13. Edwin Nelmes, *Operations Group Chairman's Factual Report*, National Transportation Safety Board, SA-400, Exhibit No. 2A, 1967, 1.
 14. *Ibid.*, 26.
 15. *Ibid.*
 16. *Ibid.*

Chapter 13

1. Mike Monroney, Hearing Before the Committee on Commerce, United States Senate, Ninetieth Congress, First Session, on Nomination of Alan S. Boys, of Florida, to be Secretary of Transportation, Warren Magnuson, Chairman (Washington, D.C.: Government Printing Office, 1967), 18.
2. Ibid.
3. Ibid., 2.
4. Richard Kent, *Safe, Separated and Soaring: A History of Federal Civil Aviation Policy 1961–1972* (Washington, D.C.: U.S. Department of Transportation, Federal Aviation Administration, 1980), 212.
5. Ibid., 213.
6. Ibid., 215.
7. Ibid.
8. Hearing Before the Committee on Interstate and Foreign Commerce, House of Representatives, Ninetieth Congress, First Session, on Regulatory Agencies under Jurisdiction of the Committee (Washington, D.C.: Government Printing Office, 1967), 400.
9. Ibid.
10. Ibid., 401.
11. Ibid.
12. *New York Times*, July 28, 1967.
13. Ibid.
14. Ibid.
15. Ibid.
16. Ibid.
17. Ibid.
18. Lyndon B. Johnson, "Text of Johnson's Statement on Status of Nation's Defenses and Race for Space," *New York Times*, January 8, 1958, 10.
19. Ibid.
20. Ibid.
21. Lyndon B. Johnson, *Memorandum for the President, Evaluation of Space Program*, April 28, 1961), 2.
22. Ibid.
23. Ibid.
24. Ibid., 7.
25. Loyd S. Swenson, Jr., James M. Grimwood, and Charles C. Alexander, *This New Ocean, A History of Project Mercury* (Washington, D.C.: National Aeronautics and Space Administration, 1998), 134.
26. Ibid., 508.
27. James M. Grimwood and Barton C. Hacker with Peter J. Vorzimmer, *Project Gemini, Technology and Operations* (Washington, D.C.: National Aeronautics and Space Administration, 1969), v.
28. Ibid., 283.
29. Courtney G. Brooks, James M. Grimwood, and Loyd D. Swinson, Jr., *Chariots For Apollo: A History of Manned Lunar Spacecraft* (Washington, D.C.: National Aeronautics and Space Administration, 1979), xiii.
30. Apollo 204 Accident, "Report of the Committee on Aeronautical and Space Sciences, United States Senate with Additional Views," Clinton P. Anderson, Chairman (Washington, D.C.: U.S. Government Printing Office, 1968), 5.
31. Ibid., 6.
32. Ibid., 8.
33. Ibid., 9.

Chapter 14

1. William McKee, *House of Representatives, Committee on Interstate and Foreign Commerce*, Harley Staggers, Chairman (Washington, D.C.: Government Printing Office, 1968), 22.
2. Charles Ruby, *House of Representatives, Subcommittee on Transportation and Aeronautics*, Samuel N. Friedel, Chairman (Washington, D.C.: Government Printing Office, 1968), 185.
3. Ibid.
4. Ibid.
5. Ibid.
6. Ibid.
7. Ibid.
8. Ibid.
9. Ibid.
10. John Robert Reagan, Deposition Held at the Federal Courthouse, Western District of North Carolina, March 19, 1969, 3166.
11. Ibid.
12. John D. Rawson, *Group Chairman's Factual Report of Investigation*, National Transportation Safety Board, SA-400, Exhibit No. 11A, 1967, 10.
13. Stuart Tipton, *House of Representatives, Subcommittee on Transportation and Aeronautics*, Samuel N. Friedel, Chairman (Washington, D.C.: Government Printing Office, 1968), 225.
14. Ibid., 226.
15. Ibid.
16. Ibid., 227.
17. Ibid.
18. Ibid.
19. Ibid., 227–228.
20. Ibid., 228.
21. Ibid., 230.
22. NAGE, http://www.nage.org/history.shtml, accessed June 23, 2012.
23. Clifford P. Burton, *House of Representatives, Subcommittee on Transportation and Aeronautics*, Samuel N. Friedel, Chairman (Washington, D.C.: Government Printing Office, 1968), 305.
24. F. Lee Bailey, *House of Representatives, Subcommittee on Transportation and Aeronautics*,

Samuel N. Friedel, Chairman (Washington, D.C.: Government Printing Office, 1968), 352.
25. Stanley Lyman, *House of Representatives, Subcommittee on Transportation and Aeronautics*, Samuel N. Friedel, Chairman (Washington, D.C.: Government Printing Office, 1968), 290–291.
26. *Ibid.*, 283.
27. *Ibid.*
28. *Ibid.*
29. *Ibid.*, 285.
30. *Ibid.*, 285.
31. *Ibid.*, 286.
32. *Ibid.*
33. *Ibid.*
34. Lyndon B. Johnson, Letter to the Administrator, Federal Aviation Agency, Commending the Agency's Record in Cost Reduction, Safety and Service, July 6, 1966.
35. *Ibid.*
36. Stanley Lyman, *House of Representatives, Subcommittee on Transportation and Aeronautics*, Samuel N. Friedel, Chairman (Washington, D.C.: Government Printing Office, 1968), 287.
37. *Ibid.*, 291.
38. *Ibid.*
39. *Ibid.*
40. *Ibid.*, 292.
41. *Ibid.*
42. *Ibid.*
43. *Ibid.*
44. *Ibid.*
45. *Ibid.*
46. *Ibid.*
47. *Ibid.*
48. *Ibid.*, 297.
49. *Ibid.*
50. *Ibid.*, 298.
51. *Ibid.*
52. *Ibid.*
53. *Ibid.*
54. *Ibid.*
55. *Ibid.*
56. *Ibid.*, 299.
57. *Ibid.*
58. *Ibid.*
59. *Ibid.*
60. *Ibid.*
61. *Ibid.*, 299–300.

Chapter 15

1. *Biography of Drew Pearson* by Jim Heintze, www.library.american.edu/pearson/biography.html, accessed July 4, 2012.
2. William Manchester, *American Caesar* (New York: Doubleday, 1968), 171.
3. Jack Anderson, "Brother Act," Bell-McClure Syndicate, August 1, 1967. Accessed from American University Archives & Special Collections.
4. *Ibid.*
5. *Ibid.*
6. *Ibid.*
7. *Ibid.*
8. *Ibid.*
9. *Ibid.*
10. *Ibid.*

Chapter 16

1. John D. Rawson, *Group Chairman's Factual Report of Investigation*, National Transportation Safety Board, SA-400, Exhibit No. 11A, 1967, 1–2.
2. *Ibid.*, 2.
3. *Ibid.*
4. *Ibid.*, 3.
5. *Ibid.*
6. *Ibid.*, 4.
7. *Ibid.*
8. *Ibid.*, 6.
9. *Ibid.*, 13.
10. *Ibid.*, 14.
11. John Robert Reagan, Deposition Held at the Federal Courthouse, Western District of North Carolina, March 15, 1969, 2885.
12. *Ibid.*
13. John D. Rawson, *Group Chairman's Factual Report of Investigation*, National Transportation Safety Board, SA-400, Exhibit No. 11A, 1967, 7.
14. *Ibid.*, 1.
15. Frank C. Nicholson, Deposition, Western District of North Carolina, May 1, 1969, 4657.
16. *Ibid.*, 4786.
17. *Ibid.*, 4664.
18. *Ibid.*, 4801.
19. Aircraft Accident Report, "American Airlines, Inc., Boeing 727, N1996, Near the Greater Cincinnati Airport, Constance, Kentucky, November 8, 1965" (Washington, D.C.: Civil Aeronautics Board, SA-387, File No. 1-0031, 1966), 25.
20. Frank C. Nicholson, Deposition, Western District of North Carolina, May 1, 1969, 4703.
21. *Ibid.*, 4738.
22. *Ibid.*
23. *Ibid.*
24. *Ibid.*
25. *Ibid.*
26. *Ibid.*
27. *Ibid.*, 4739.
28. *Ibid.*
29. *Ibid.*

30. Aircraft Accident Report, "American Airlines, Inc., Boeing 727, N1996, Near the Greater Cincinnati Airport, Constance, Kentucky, November 8, 1965" (Washington, D.C.: Civil Aeronautics Board, SA-387, File No. 1–0031, 1966), 25.
31. *Ibid.*
32. Frank C. Nicholson, Deposition, Western District of North Carolina, May 1, 1969, 4703.
33. Frank C. Nicholson, Deposition, Western District of North Carolina, April 29, 1969, 4657.
34. *Ibid.*, 4801.
35. *Ibid.*, 4787–4788.
36. *Ibid.*, 4796.
37. *Ibid.*, 4767.
38. *Ibid.*, 4768.
39. *Ibid.*
40. *Ibid.*, 4803.
41. *Ibid.*, 4804.
42. *Ibid.*
43. *Ibid.*
44. *Ibid.*, 4806.
45. *Ibid.*, 4807.
46. *Ibid.*
47. *Ibid.*
48. *Ibid.*, 4809.
49. *Ibid.*, 4811.
50. *Ibid.*, 4812.
51. *Ibid.*, 4815.
52. *Ibid.*, 4816.
53. *Ibid.*
54. Edwin Nelmes, *Operations Group Chairman's Factual Report*, National Transportation Safety Board, SA-400, Exhibit No. 2A, 1967, 14.

Chapter 17

1. Francis X. Graves, *ATC Group Chairman's Factual Report*, National Transportation Safety Board, SA-400, Exhibit No. 3A, 1967, 4.
2. Aircraft Accident Report, "Piedmont Aviation, Inc. Piedmont Airlines Division Boeing 727, N68650, Lanseair Inc., Cessna 310, N3121S Mid Air Collision, Hendersonville, North Carolina, July 19, 1967," (Washington, D.C.: National Transportation Safety Board, SA-400, File 1–0005, 1968), 30.
3. Francis X. Graves, *ATC Group Chairman's Factual Report*, National Transportation Safety Board, SA-400, Exhibit No. 3A, 1967, 5.
4. Aircraft Accident Report, "Piedmont Aviation, Inc. Piedmont Airlines Division Boeing 727, N68650, Lanseair Inc., Cessna 310, N3121S Mid Air Collision, Hendersonville, North Carolina, July 19, 1967," (Washington, D.C.: National Transportation Safety Board, SA-400, File 1–0005, 1968), 44.
5. T.H. Davis, "Conclusions and Recommendations of Piedmont Aviation, Inc. from Evidence Presented to National Transportation Safety Board Concerning Mid-Air Collision, Hendersonville, North Carolina, July 19, 1967," December 18, 1967, 4.
6. Federal Aviation Administration, *Transcription of Recorder Channel Number Two Approach Control/Flight Data Position*, Airport Traffic Control Tower, Asheville, North Carolina, July 19, 1967, 1558:21 (time).
7. *Ibid.*
8. Francis X. Graves, Deposition Held at the Federal Courthouse, Western District of North Carolina, January 6, 1969, 5868.
9. Federal Bureau of Investigation, "Request for Examination of Tape Recording of Transmissions between Asheville, North Carolina, Control Tower and Cessna 310 plane which Crashed along with Piedmont Airlines on July 19, 1967," FBI File No. 149–4971, Lab. No. E-481 DX, September 27, 1967.
10. U.S. Department of Commerce, National Bureau of Standards. Letter from Mr. Edwin D. Burnett, Physicist, Sound Section, Institute for Basic Standards, to Mr. Bernie Curtis, FAA. (October 5, 1967).
11. *Ibid.*
12. Francis X. Graves, *Addendum to ATC Group Chairman's Factual Report*, National Transportation Safety Board, DCA 68-A-1, 1967.
13. *Ibid.*
14. Francis X. Graves, *Cockpit Voice Recorder Specialist's Factual Report*, National Transportation Safety Board, SA-400, Exhibit No. 12A, 1967, 1.
15. Aircraft Accident Report, "North Central Airlines, Inc. Convair 580, N46345 Home Airmotive, Inc., Cessna 150, N87425, Midair Collision Near Milwaukee, Wisconsin, August 4, 1968" (Washington, D.C.: National Transportation Safety Board, SA-405, File 1–0011, 1969), 40.
16. *Ibid.*
17. W.O. Tadlock, Director of Flight Operations, "Operations Letter, Piedmont Airlines To All Pilots" (Piedmont Airlines, Letter #12–261, February 28, 1961).
18. W.O. Tadlock, Director of Flight Operations, "Operations Letter, Piedmont Airlines To All Pilots" (Piedmont Airlines, Letter #7–1160, Revised April 25, 1962).
19. Aircraft Accident Report, "American Airlines, Inc., Boeing 727, N1996, Near the Greater Cincinnati Airport, Constance, Kentucky, November 8, 1965" (Washington, D.C.:

Civil Aeronautics Board, SA-387, File No. 1–0031, 1966), 25.
20. *Ibid.*
21. *Ibid.*
22. National Transportation Safety Board, *Complete Transcript of the Last 18 Minutes of the Cockpit Voice Recorder Tape Recording Aboard Piedmont Boeing 727 Aircraft N68650*, July 19, 1967, 17.
23. *Ibid.*
24. *Ibid.*
25. *Ibid.*
26. *Ibid.*, 18.
27. *Ibid.*
28. *Ibid.*
29. "An Aircraft Accident Involving Piedmont Airlines, Inc., Boeing 727, N68650 and Lanseair Cessna 310, N3121S at Hendersonville, North Carolina, July 19, 1967," public hearing (Washington, D.C.: Department of Transportation, Docket No. SA-400, 1967), 628.
30. *Ibid.*
31. *Ibid.*, 634.
32. Aircraft Accident Report, "Piedmont Aviation, Inc. Piedmont Airlines Division Boeing 727, N68650, Lanseair Inc., Cessna 310, N3121S Mid Air Collision, Hendersonville, North Carolina, July 19, 1967" (Washington, D.C.: National Transportation Safety Board, SA-400, File 1–0005, 1968), 19.

Chapter 18

1. Asheville Regional Airport http://flyavl.com/pages/about-the-airport/general-invo/history.php (accessed September 16, 2012).
2. *Ibid.*
3. Federal Aviation Regulation, Section 91.87 "Operation at airports with operating control towers," (f)(1).
4. Earl E. Cato, chief, CS/T, Asheville, NC letter to Chief, ARTC Center, Atlanta, Georgia (September 28, 1961).
5. *Ibid.*
6. *Ibid.*
7. Federal Aviation Agency, Flight Standards Service, Standard Instrument Approach Procedure, AVL, January 13, 1962.
8. National Transportation Safety Board, *Complete Transcript of the Last 18 Minutes of the Cockpit Voice Recorder Tape Recording Aboard Piedmont Boeing 727 Aircraft N68650*, July 19, 1967, 8.
9. *Ibid.*, 11.
10. *Ibid.*
11. *Ibid.*, 12.
12. Roys C. Jones, "Analysis of Air Traffic Control Exercised on N3121S, A Cessna 310 Aircraft and Piedmont Flight 22, A Boeing 727 Aircraft on July 19, 1967, Prior to Their Mid-Air Collision Near Hendersonville, North Carolina at 1600 Hours, One Minute and Sixteen Seconds, GMT" (August 10, 1967, photocopied), 7.
13. National Transportation Safety Board, *Complete Transcript of the Last 18 Minutes of the Cockpit Voice Recorder Tape Recording Aboard Piedmont Boeing 727 Aircraft N68650*, July 19, 1967, 16.

Chapter 19

1. Air Safety Investigator's Manual, Bureau of Safety, Civil Aeronautics board, January 31, 1964, Section 2.21, 3.
2. *Ibid.*, 4.
3. *Ibid.*
4. Stephen Barlay, *Air Crash Detective: An International Report on the Quest for Air Safety* (London: Hamish Hamilton, 1969), 182–183.
5. Jack Anderson, "Brother Act," Bell-McClure Syndicate, August 1, 1967.
6. Harold Kimball Saunders, Sr., Affidavit in Support of US Airways [Successor in interest to Piedmont Airlines] Response to Petition of Modification filed March 2005 by Paul D. Houle (Forsyth County, North Carolina, September 5, 2006).
7. Air Safety Investigator's Manual, Bureau of Safety, Civil Aeronautics Board, January 31, 1964, Section 2.21, 3.
8. *Ibid.*, 4.
9. Cynthia C. Lebow, Liam P. Sarsfield, William L. Stanley, Emile Ettedgui, and Garth Henning, *Safety in the Skies: Personnel and Parties in NTSB Aviation Accident Investigations* (Institute for Civil Justice, RAND, 2000), 7.
10. John Robert Reagan, Deposition Held at the Federal Courthouse, Western District of North Carolina, March 19, 1969, 3083.
11. *Ibid.*
12. *Ibid.*, 3516.
13. Frank C. Nicholson, Deposition, Western District of North Carolina, April 29, 1969, 4524–4525.
14. *Ibid.*, 4525.
15. *Ibid.*, 4647.
16. Prater Hogue, Letter to Zeke Saunders, August 8, 1967, file number 6–7/11–1334-I/C.

Chapter 20

1. *New York Times*, September 18, 1967.
2. *Ibid.*
3. *Ibid.*
4. Charles Schultze, http://en.wikipedia.

org/wiki/Charles_Schultze, accessed October 28, 2012.
5. Memo, Charles Schultze to the President, 9/18/1967, WHCF, LBJ Library.
6. *Ibid.*
7. *Ibid.*
8. *Ibid.*
9. *Ibid.*
10. *Ibid.*
11. *Ibid.*
12. *Ibid.*
13. *Ibid.*
14. *Ibid.*
15. Richard J. Kent, Jr., *Safe, Separated, and Soaring: A History of Federal Civil Aviation Policy 1961–1972* (Washington, D.C.: U.S. Department of Transportation, Federal Aviation Administration, 1980), 201.
16. Memo, Joe Laitin to George Christian, 9/19/67, WHCF, LBJ Library.
17. Transcript of News Conference with George Christian, September 20, 1967, 11:15 a.m., LBJ Library.
18. Letter, President Johnson to Alan Boyd, 9/20/67, WHCF, LBJ Library.
19. *Ibid.*
20. *Ibid.*
21. *Ibid.*

Chapter 21

1. Letter from Civil Aeronautics Board to Elwood R. Quesada, June 1962.
2. *Ibid.*
3. "An Aircraft Accident Involving Piedmont Airlines, Inc., Boeing 727, N68650 and Lanseair Cessna 310, N3121S at Hendersonville, North Carolina, July 19, 1967," public hearing (Washington, D.C.: Department of Transportation, Docket No. SA-400, 1967), 600.
4. *Ibid.*
5. *Ibid.*, 601.
6. *Winston-Salem Journal,* October 11, 1967.
7. "An Aircraft Accident Involving Piedmont Airlines, Inc., Boeing 727, N68650 and Lanseair Cessna 310, N3121S at Hendersonville, North Carolina, July 19, 1967," public hearing (Washington, D.C.: Department of Transportation, Docket No. SA-400, 1967), 150.
8. *Ibid.*, 152.
9. *The Times-News* (North Carolina), October 11, 1967.
10. *Ibid.*
11. "An Aircraft Accident Involving Piedmont Airlines, Inc., Boeing 727, N68650 and Lanseair Cessna 310, N3121S at Hendersonville, North Carolina, July 19, 1967," (Washington, D.C.: Department of Transportation, Docket No. SA-400, 1967), 159.
12. *Ibid.*, 628.
13. *Ibid.*
14. *Ibid.*
15. Aircraft Accident Report, "American Airlines, Inc., Boeing 727, N1996, Near the Greater Cincinnati Airport, Constance, Kentucky, November 8, 1965" (Washington, D.C.: Civil Aeronautics Board, File No. 1–0031, 1966), 25.
16. *Ibid.*
17. Aircraft Accident Report, "Piedmont Aviation, Inc. Piedmont Airlines Division Boeing 727, N68650, Lanseair Inc., Cessna 310, N3121S Mid Air Collision, Hendersonville, North Carolina, July 19, 1967" (Washington, D.C.: National Transportation Safety Board, SA-400, File 1–0005, 1968), 2.
18. Francis McAdams, *Notation 120-A Aircraft accident report—Piedmont Airlines, Inc., Boeing 727/Lanseair Inc., Cessna 310 N3121S midair collision,, Hendersonville, North Carolina, July 19, 1967* (Washington, D.C.: Department of Transportation, National Transportation Safety Board, November 15, 1968), 1.
19. *Ibid.*
20. *Ibid.*
21. C.O. Miller, *The Bureau of Aviation Safety and the Accident Prevention System,* remarks at the FSF Luncheon, Anaheim, CA, October 9, 1968, 6.
22. *Ibid.*
23. C.O. Miller, *Board Presentation re Probable Cause-Hendersonville Case,* handwritten notes, November 20, 1968.
24. C.O. Miller and J. O'Connell, *The Bureau of Aviation Safety and the Accident Prevention System,* Remarks at the FSF Luncheon, Anaheim, CA, October 9, 1968.
25. *Ibid.*
26. Francis McAdams, *Notation 120-A Aircraft accident report—Piedmont Airlines, Inc., Boeing 727/Lanseair Inc., Cessna 310 N3121S midair collision,, Hendersonville, North Carolina, July 19, 1967* (Washington, D.C.: Department of Transportation, National Transportation Safety Board, November 15, 1968), 1–2.
27. Aircraft Accident Report, "Pacific Southwest Airlines, Inc., B-727, N533PS and Gibbs Flite Center, Inc., Cessna 172, N7711G, San Diego, California, September 25, 1978" (Washington, D.C.: National Transportation Safety Board, NTSB-AAR 79-5 1979), 38.
28. Memo, Richard Woody, Meteorologist, WBAS, Asheville, NC, to Director, Weather Bureau, Silver Spring, Maryland, 7/24/1967, 652.13, United States Government.
29. *Ibid.*

30. J.J. Barnes, Interview by the National Transportation Safety Board, "Witness Statement of Mid-Air Collision," Hendersonville, North Carolina, August 7, 1967.
31. Aircraft Accident Report, "Piedmont Aviation, Inc. Piedmont Airlines Division Boeing 727, N68650, Lanseair Inc., Cessna 310, N3121S Mid Air Collision, Hendersonville, North Carolina, July 19, 1967" (Washington, D.C.: National Transportation Safety Board, SA-400, File 1–0005, 1968), 37.
32. Russell J. Abbott, *Group Chairman's Factual Report of Investigation*, National Transportation Safety Board, SA-400, Exhibit No. 4A, 1967, 6.
33. Aircraft Accident Report, "Piedmont Aviation, Inc. Piedmont Airlines Division Boeing 727, N68650, Lanseair Inc., Cessna 310, N3121S Mid Air Collision, Hendersonville, North Carolina, July 19, 1967" (Washington, D.C.: National Transportation Safety Board, SA-400, File 1–0005, 1968), 7.
34. *Ibid.*, Appendix A, i.
35. *Ibid.*, 19.
36. *Risk Management Handbook.* Department of Transportation, Federal Aviation Administration, Flight Standards Service, 2009, Chapter 1, 3.
37. Aircraft Accident Report, "Piedmont Aviation, Inc. Piedmont Airlines Division Boeing 727, N68650, Lanseair Inc., Cessna 310, N3121S Mid Air Collision, Hendersonville, North Carolina, July 19, 1967" (Washington, D.C.: National Transportation Safety Board, SA-400, File 1–0005, 1968), 13.
38. *Ibid.*, 36.
39. *Ibid.*, 36–37.
40. *Ibid.*, 29–30.
41. *Ibid.*, 31.
42. *Ibid.*, 39.
43. Francis McAdams, *Notation 120-A Aircraft accident report—Piedmont Airlines, Inc., Boeing 727/Lanseair Inc., Cessna 310 N3121S midair collision, Hendersonville, North Carolina, July 19, 1967* (Washington, D.C.: Department of Transportation, National Transportation Safety Board, November 15, 1968), 2.
44. *Ibid.*
45. *Ibid.*, 3.
46. *Ibid.*
47. *Ibid.*
48. *Ibid.*
49. Aircraft Accident Report, "Piedmont Aviation, Inc. Piedmont Airlines Division Boeing 727, N68650, Lanseair Inc., Cessna 310, N3121S Mid Air Collision, Hendersonville, North Carolina, July 19, 1967" (Washington, D.C.: National Transportation Safety Board, SA-400, File 1–0005, 1968), 5.
50. Francis X. Graves, *ATC Group Chairman's Factual Report*, National Transportation Safety Board, SA-400, Exhibit No. 3A, 1967, 13.

Chapter 22

1. Aircraft Accident Report, "Eastern Air Lines, Inc., Douglas DC-9–31, N8984E, Charlotte, North Carolina, September 11, 1974" (Washington, D.C.: National Transportation Safety Board, File 1–0020, 1975), 1.
2. "Safety Recommendations 74–85 & 74–86," to Honorable Alexander P. Butterfield, Administrator, FAA, October 8, 1974.
3. Robert Sumwalt, "The Sterile Cockpit," *ASRS DirectLine*, Issue #4, June 1993.
4. *Ibid.*
5. Aircraft Accident Report, "Pacific Southwest Airlines, Inc., B727-N533PS and Gibbs Flite Center, Inc., Cessna 172, N7711G, San Diego, California September 25, 1978" (National Transportation Safety Board: Washington D.C. Report No. 79–5, 1979), 1.
6. *Ibid.*, 31–32.
7. *Ibid.*, 32.
8. *Ibid.*

Chapter 23

1. Acting Director, Office of Aviation Safety, Memorandum "Response to Petition for Reconsideration regarding an aviation accident involving a Boeing 727, N68650, and a Lanseair Cessna 310, N3121S, near Hendersonville, North Carolina, on July 19, 1967," Aircraft Accident No. DCA68A000, February 9, 2007.
2. Aircraft Accident Report, "Piedmont Aviation, Inc. Piedmont Airlines Division Boeing 727, N68650, Lanseair Inc., Cessna 310, N3121S Mid Air Collision, Hendersonville, North Carolina, July 19, 1967" (Washington, D.C.: National Transportation Safety Board, SA-400, File 1–0005, 1968), 30.
3. Acting Director, Office of Aviation Safety, Memorandum "Response to Petition for Reconsideration regarding an aviation accident involving a Boeing 727, N68650, and a Lanseair Cessna 310, N3121S, near Hendersonville, North Carolina, on July 19, 1967," Aircraft Accident No. DCA68A000, February 9, 2007, 3.
4. Francis McAdams, *Notation 120-A Aircraft accident report—Piedmont Airlines, Inc., Boeing 727/Lanseair Inc., Cessna 310 N3121S midair collision,, Hendersonville, North Carolina, July 19, 1967* (Washington, D.C.: Department of Transportation, National Transportation Safety Board, November 15, 1968), 3.

5. Acting Director, Office of Aviation Safety, Memorandum "Response to Petition for Reconsideration regarding an aviation accident involving a Boeing 727, N68650, and a Lanseair Cessna 310, N3121S, near Hendersonville, North Carolina, on July 19, 1967," Aircraft Accident No. DCA68A000, February 9, 2007, 4.

6. *Ibid.*

7. Harold Kimball Saunders, Sr., Affidavit in Support of US Airways [Successor in interest to Piedmont Airlines] Response to Petition of Modification filed March 2005 by Paul D. Houle (Forsyth County, North Carolina, September 5, 2006).

8. Acting Director, Office of Aviation Safety, Memorandum "Response to Petition for Reconsideration regarding an aviation accident involving a Boeing 727, N68650, and a Lanseair Cessna 310, N3121S, near Hendersonville, North Carolina, on July 19, 1967," Aircraft Accident No. DCA68A000, February 9, 2007, 4.

9. *Ibid.*

BIBLIOGRAPHY

Abbott, Russell J. *Group Chairman's Factual Report of Investigation.* National Transportation Safety Board, 1967.
Acting Director, Office of Aviation Safety. "Response to Petition for Reconsideration regarding an aviation accident involving a Boeing 727, N68650, and a Lanseair Cessna 310, N3121S, near Hendersonville, NC, on July 19, 1967. Aircraft Accident No. DCA68A000." Memorandum, February 9, 2007.
"Air Crash Statements 'Unfair,' Says Dando." *Springfield Leader & Press*, July 24, 1967.
Air Safety Investigator's Manual. Bureau of Safety, Civil Aeronautics Board, January 31, 1964.
"Air Safety Steps for Rising Traffic Ordered by FAA." *New York Times*, May 13, 1967.
"Air Traffic Aide Hits Economies." *New York Times*, February 6, 1966.
"Air Traffic Group Hits Mental Test." *New York Times*, April 30, 1967.
"An Aircraft Accident Involving Piedmont Airlines, Inc., Boeing 727, N68650 and Lanseair Cessna 310, N3121S at Hendersonville, NC, July 19, 1967." Public hearing, Department of Transportation, Washington, D.C., 1967.
Allen, Bobbie. *House of Representatives, Committee on Interstate and Foreign Commerce.* Washington, D.C.: Government Printing Office, 1968.
"American Airlines, Inc., Boeing 727, N1996, Near the Greater Cincinnati Airport, Constance, Kentucky, November 8, 1965." Aircraft Accident Report. Washington, D.C.: Civil Aeronautics Board, 1966.
Anderson, Jack. "Brother Act." Bell-McClure Syndicate, August 1, 1967.
"Asheville Regional Airport." http://flyavl.com/pages/about-the-airport/general-invo/history.php. Accessed September 16, 2012.
Bailey, F. Lee. *House of Representatives, Subcommittee on Transportation and Aeronautics.* Washington, D.C.: Government Printing Office, 1968.
Barlay, Stephen. *Air Crash Detective: An International Report on the Quest for Air Safety.* London: Hamish Hamilton, 1969.
Barnes, J. J. "Witness Statement of Mid-Air Collision." Hendersonville, NC, August 7, 1967.
Beddingfield, Roy. "Witness Statement of Mid-Air Collision." Hendersonville, NC, July 21, 1967.
Benham, Lance. Author interview, Hendersonville, NC, 2004.
Bedwell, Don. "Going Nowhere Fast." *American Heritage*, Winter 2006, Volume 21.
Borsky, Paul N. *Community Reactions to Sonic booms, Oklahoma City Area, February-July 1964, Part 1, Major Findings.* National Opinion Research Center, University of Chicago, 1965.
"Braniff Airlines, Inc., Lockheed Electra, N9705C, Buffalo, Texas, September 29, 1959." Aircraft Accident Report. Washington, D.C.: Civil Aeronautics Board, 1961.
Brooks, Courtney G., James M. Grimwood, and Loyd D. Swinson Jr. *Chariots for Apollo: A History of Manned Lunar Spacecraft.* Washington, D.C.: National Aeronautics and Space Administration, 1979.

Burnett, Edwin. "Letter to Mr. Bernie Curtis, FAA." U.S. Department of Commerce, National Bureau of Standards, Sound Section, Institute for Basic Standards, October 5, 1967.
Burton, Clifford. *House of Representatives, Subcommittee on Transportation and Aeronautics.* Washington, D.C.: Government Printing Office, 1968.
"C.A.B Finds No Reason to Ground Boeing 727's." *New York Times*, November 14, 1965.
"Campers Stand Up to an Emergency." *Atlanta Constitution*, July 20, 1967.
Case, Holly Anderson. Author interview, Hendersonville, NC, 2004.
Cato, Earl E. "Letter to Chief, ARTC Center, Atlanta, Georgia." September 28, 1961.
"Cessna Had Direction Finder on Wrong Beacon." *The Times-News*, October 11, 1967.
"Charles Schultze." http://en.wikipedia.org/wiki/Charles_Schultze. Accessed October 28, 2012.
Chase, Nan. *Asheville: A History.* Jefferson, NC: McFarland, 2007.
Childs, James T. *Structures Group Chairman's Factual Report.* National Transportation Safety Board, 1967.
Christian, George. Transcript of News Conference, September 20, 1967.
Clark, Evert. "F.A.A. is Pressing for More Funds in Safety Drive." *New York Times*, September 18, 1967.
Complete Transcript of the last 18 minutes of the Cockpit Voice Recorder Tape Recording Aboard Piedmont Boeing 727 Aircraft N68650. National Transportation Safety Board, 1967.
"Congressmen Urge Limiting Small Planes." *Winston-Salem Journal*, July 21, 1967.
"Controllers Union Urges M'Kee Ouster." *New York Times*, July 28, 1967.
Cromley, Allan. "Monroney's Signal Loud, Clear: No More SST Booms for Oklahoma." *The Oklahoman*, February 20, 1966.
Davis, Andrew J. *Weather Conditions in Southeastern United States*, 1967.
Davis, Verne R. "Witness Statement of Mid-Air Collision." Hendersonville, NC, July 23, 1967.
Dermid, Joel. "Witness Statement of Mid-Air Collision." Hendersonville, NC, July 21, 1967.
Dunn, J.A.C. "The History of Piedmont Airlines." *Pace*, December 1988.
Dwiggins, Don. *The SST: Here it Comes, Ready or Not.* New York: Doubleday, 1968.
"Eastern Airlines Inc., Douglas DC-8, N8607, New Orleans, Louisiana, February 25, 1964." Aircraft Accident Report. Washington, D.C.: Civil Aeronautics Board, 1966.
"Eastern Air Lines, Inc., Douglas DC-9-31, N8984E, Charlotte, NC, September 11, 1974," Aircraft Accident Report. Washington, D.C.: National Transportation Safety Board, 1975.
Federal Aviation Administration. "Operation at airports with operating control towers," *Federal Aviation Regulations*. Section 91.
Federal Aviation Agency. *Federal Aviation Regulations.* 14 CFR, Section 91.87, 1967.
Federal Aviation Agency. *Federal Aviation Regulations.* 14 CFR, Section 121.357. December 31, 1964.
Federal Aviation Agency, "Standard Instrument Approach Procedure." AVL. Flight Standards Service, January 13, 1962.
Federal Aviation Agency. *Standard Instrument Approach Procedure-Type ILS.* Asheville Municipal Airport, 16 November 1963.
Federal Bureau of Investigation, "Request for Examination of Tape Recording of Transmissions between Asheville, NC, Control Tower and Cessna 310 plane which Crashed along with Piedmont Airlines on July 19, 1967." FBI File No. 149–4971, September 27, 1967.
Folk, Benjamin M. SATCS, Narrative Statement, 1967.
Gillilland, Whitney. Speech given at the University of Miami. Iowa City: University of Iowa, 1968.
Graves, Francis, X. *Addendum to ATC Group Chairman's Factual Report.* National Transportation Safety Board, 1967.

Graves, Francis X. *ATC Group Chairman's Factual Report*. National Transportation Safety Board, 1967.
Griffin, Lacy. Personal conversation, Hendersonville, NC, 2000.
Grimwood, James M., Barton C. Hacker, and Peter Vorzimme. *Project Gemini: Technology and Operations*. Washington, D.C.: National Aeronautics and Space Administration, 1969.
Harrison, Benjamin T. and Christopher L. Mosher, "John T. McNaughton and Vietnam: The Early Years as Assistant Secretary of Defense, 1964–1965." *History: The Journal of the Historical Association*, October 2007.
Hearings before the Committee on Commerce, United States Senate, Ninetieth Congress, First Session, on Nominations of Joseph P. O'Connell, Francis H. McAdams, Rear Adm. Louis N. Thayer, John H. Reed, Oscar Laurel. Washington, D.C.: Government Printing Office, 1967.
Hearing before the Committee on Interstate and Foreign Commerce, House of Representatives, Ninetieth Congress, First Session, on Regulatory Agencies under Jurisdiction of the Committee. Washington, D.C.: Government Printing Office, 1967.
Heintze, Jim. "Biography of Drew Pearson." www.library.american.edu/pearson t /biography.html. Accessed July 4, 2012.
Herring, George C., ed. *The Pentagon Papers*. New York: McGraw-Hill, 1993.
"Historic Touchdown." *The Times News*, January 16, 1961.
Hoar, Steve. "Downed Jet was Plagued by Troubles." *Winston-Salem Journal*, July 20, 1967.
Hogue, Prater. "Letter to Zeke Saunders." August 8, 1967.
Hudson, Edward. "Airlines to meet on 727 Jet Liner; Conference called by F.A.A., for tomorrow in Capital." *New York Times*, February 16, 1966.
Hudson, Edward. "Changes Sought for 727 Jet crews; F.A.A. Finds no flaws in design of Boeing Craft." *New York Times*, February 18, 1966.
Hurley, Richard. "The Passing of the Pacemaker." *Airliners*, 1988.
"Jet Pilot Within Regulations on Takeoff, Witness Asserts." *Winston-Salem Journal*, October 11, 1967.
Johnson, Lyndon, B. Letter to Alan Boyd, September 20, 1967.
Johnson, Lyndon, B. "Letter to the Administrator, Federal Aviation Agency, Commending the Agency's Record in Cost Reduction, Safety and Service." July 6, 1966.
Johnson, Lyndon B. "Memorandum for The President, Evaluation of Space Program." Washington, D.C.: Office of the Vice President, April 28, 1961.
Johnson, Lyndon B. "Remarks at the Swearing in of General McKee as Administrator, Federal Aviation Agency." July 1, 1965.
Johnson, Lyndon B. "Statement by the President on the Death of John T. McNaughton," July 19, 1967.
Johnson, Lyndon B. "Text of Johnson's Statement on Status of Nation's Defenses and Race for Space." *New York Times*, January 8, 1958.
Johnson, Lyndon B. The President's News Conference, Austin, Texas. December 6, 1966.
Johnson, Lyndon. B. The President's News Conference at the LBJ Ranch. December 31, 1966.
Jones, Roys. *Analysis of Air Traffic Control Exercised on N3121S, A Cessna 310 Aircraft and Piedmont Flight 22, A Boeing 727 Aircraft on July 19, 1967, Prior to Their Mid-Air Collision Near Hendersonville, North Carolina at 1600 Hours, One Minute and Sixteen Seconds, GMT*, 1967.
Kapustin, Rudolf. *Power Plant Group Chairman's Factual Report*. National Transportation Safety Board, 1967.
Kaufman, Robert. Personal telephone conversation, Miami, Florida.
Kayne, Victor. *House of Representatives, Subcommittee on Transportation and Aeronautics, Committee on Interstate and Foreign Commerce*. Washington, D.C.: Government Printing Office, 1968.

Kennedy, John F. "Remarks at Colorado Springs to the Graduating Class of the U.S. Air Force Academy," June 5, 1963.
Kent, Richard. *Safe, Separated and Soaring, A History of Federal Civil Aviation Policy 1961–1972*. Washington, D.C.: U.S. Department of Transportation, Federal Aviation Administration, 1980.
Laitin, Joe. "Memo to George Christian." September 19, 1967.
Lebow, Cynthia, Liam P. Sarsfield, William L. Stanley, Emile Ettedgui, and Garth Henning. *Safety in the Skies: Personnel and Parties in NTSB Aviation Accident Investigations*. Institute for Civil Justice, RAND, 2000.
Lee, Laurence. Personal conversation, Greenville, South Carolina, 2010.
Lee, Laurence. "Witness Statement of Mid-Air Collision." Hendersonville, July 21, 1967.
Letter from the Civil Aeronautics Board to Elwood R. Quesada, June 1962.
Lyman, Stanley. *House of Representatives, Subcommittee on Transportation and Aeronautics, Committee on Interstate and Foreign Commerce*. Washington, D.C., 1968.
Manchester, William. *American Caesar*. New York: Doubleday, 1968.
McAdams, Francis. "Notation 120-A Aircraft accident report—Piedmont Airlines, Inc., Boeing 727/Lanseair Inc., Cessna 310 N3121S midair collision, Hendersonville, July 19, 1967." November 15, 1968.
McCaleb, Hubert. *Weather Group Chairman's Factual Report*. National Transportation Safety Board, 1967.
McKee, William F. Hearings Before the Committee on Commerce, United States Senate, Eighty-ninth Congress, First Session, on Nomination of Gen. William F. McKee, Administrator, Federal Aviation Agency. Washington, D.C., 1965.
McKee, William. *House of Representatives, Committee on Interstate and Foreign Commerce*. Washington, D.C.: Government Printing Office, 1968.
McNamara, Robert, and Brian VanDeMark. *In Retrospect*. New York: Times Books, 1995.
McNamara, Robert. "Memo to the President." May 19, 1967.
Miller, C.O. "Board Presentation re Probable Cause-Hendersonville case." Handwritten notes, November 20, 1968.
Miller, C.O. "The Bureau of Aviation Safety and the Accident Prevention System." Remarks at the FSF Luncheon, October 9, 1968.
Monroney, Mike. Hearing Before the Committee on Commerce, United States Senate, Ninetieth Congress, First Session, on Nomination of Alan S. Boys, of Florida, to be Secretary of Transportation. Washington, D.C.: Government Printing Office, 1967.
Murray, William. "Witness Statement of Mid-Air Collision." Hendersonville, July 21, 1967.
"NAGE." http://www.nage.org/history.shtml. Accessed June 23, 2012.
National Opinion Research Center, University of Chicago, January 1965.
National Park Service. "National Register of Historical Places Travel Itinerary, Asheville, NC." http://www.nps.gov/nr/travel/asheville/architecture.htm. Accessed August 21, 2013.
Nelmes, Edwin. *Operations Group Chairman's Factual Report*. National Transportation Safety Board, 1967.
Nicholson, Frank C. Deposition, Western District of North Carolina, 1969.
"North Central Airlines, Inc. Convair 580, N46345 Home Airmotive, Inc., Cessna 150, N87425, Midair Collision Near Milwaukee, Wisconsin, August 4, 1968." Aircraft Accident Report. Washington, D.C.: National Transportation Safety Board, 1969.
"Northwest Airlines, Inc., Boeing 720B, N724US, Near Miami, Florida, February 12, 1963." Aircraft Accident Report. Washington, D.C.: Civil Aeronautics Board, 1965.
"Northwest Airlines, Lockheed Electra, N121US, Near Cannelton, Indiana, March 17, 1960." Aircraft Accident Report. Washington, D.C.: Civil Aeronautics Board, 1961.
O'Connell, Joseph J. *House of Representatives Committee on Interstate and Foreign Commerce*. Washington, D.C.: Government Printing Office, 1968.
"One Man's Anguish." *Time*, May 5, 1961.
O'Neil, Paul O. "Brilliant Detection in Jet Age Mystery." *Life*, July 25, 1960.

Ottinger, Richard L. *House of Representatives, Committee on Interstate and Foreign Commerce*, Washington, D.C.: Government Printing Office, 1968.
Ottinger, Richard Lawrence. *Biographical Directory of the United States Congress, 1774–Present.* http://bioguide.congress.gov/scripts/biodisplay.pl?index=O000134. Accessed April 15, 2012.
"Pacific Southwest Airlines, Inc., B-727, N533PS and Gibbs Flite Center, Inc., Cessna 172, N7711G, San Diego, California, September 25, 1978." Aircraft Accident Report. Washington, D.C.: National Transportation Safety Board, 1979.
Painter, Ronald A. ATCS, Narrative Statement, 1967.
"Pan American World Airways, Inc., Boeing 707-121, N709-PA, Near Elkton, Maryland, December 8, 1963." Aircraft Accident Report. Washington, D.C.: Civil Aeronautics Board, 1965.
Patton, O.E. *Flight Recorder Chairman's Factual Report*. National Transportation Safety Board, 1967.
Pearson, Drew. "Facts on Air Crashes Are Now Suppressed." "Washington Merry-Go-Round," July 6, 1967.
"Piedmont Airlines, Douglas DC-3, N55V, On Bucks Elbow Mountain, Near Charlottesville, Virginia, October 30, 1959." Aircraft Accident Report. Washington, D.C.: Civil Aeronautics Board, 1961.
"Piedmont Aviation, Inc. Piedmont Airlines Division Boeing 727, N68650, Lanseair Inc., Cessna 310, N3121S Mid Air Collision, Hendersonville, July 19, 1967." Aircraft Accident Report. Washington, D.C.: National Transportation Safety Board, 1968.
"Plane Collisions Alarm Congress; House unit Plans Hearings in wake of Carolina Crash." *New York Times*, July 21, 1967.
Rawson, John D. *Group Chairman's Factual Report of Investigation*. National Transportation Safety Board, 1967.
Reagan, John R. Deposition, Western District of North Carolina, 1969.
"Report of the Committee on Aeronautical and Space Sciences, United States Senate, with Additional Views." Apollo 204 Accident. Washington, D.C.: U.S. Government Printing Office, 1968.
Risk Management Handbook. Department of Transportation, Federal Aviation Administration, Flight Standards Service, 2009.
Ruby, Charles. *House of Representatives, Subcommittee on Transportation and Aeronautics*. Washington, D.C.: Government Printing Office, 1968.
"Safety Board Will Conduct Official Inquiry of Crash." *Times-News*, July 20, 1967.
"Safety Recommendations 74–85 & 74–86." To Honorable Alexander P. Butterfield, Administrator, FAA, October 8, 1974.
Saunders, Harold Kimball, Sr. Affidavit in support of US Airways [Successor in interest to Piedmont Airlines] Response to Petition of Modification filed March 2005 by Paul D. Houle. Forsyth County, NC, 2006.
Schultze, Charles. "Memo to the President." September 18, 1967.
Shuler, Blanche. Author interview, Hendersonville, 2004.
"SST Price & Progress." *Time*, September 30, 1966.
Staggers, Harley. *House of Representatives, Committee on Interstate and Foreign Commerce*. Washington, D.C.: Government Printing Office, 1968.
Sumwalt, Robert. "The Sterile Cockpit," *ASRS DirectLine*, Issue #4, June 1993.
Supersonic Transport Development Program. Federal Aviation Agency, 1967.
Swenson, Loyd S., James M. Grimwood, and Charles C. Alexander. *This New Ocean: A History of Project Mercury*. Washington, D.C.: National Aeronautics and Space Administration, 1998.
Tadlock, W.O., Director of Flight Operations. "Operations Letter, Piedmont Airlines to All Pilots." Piedmont Airlines, Letter#7–1160. Revised April 25, 1962.
Tadlock, W.O., Director of Flight Operations. "Operations Letter, Piedmont Airlines to All Pilots." Piedmont Airlines, Letter#12–261. February 28, 1961.

Taylor, Sally. Author interview, Hendersonville, 2006.
Terrell, Bob. *Historic Asheville (1792–1930)*. North Carolina: WorldComm, 1997.
Tipton, Stuart. *House of Representatives, Subcommittee on Transportation and Aeronautics.* Washington, D.C.: Government Printing Office, 1968.
Transcript of Conversations Recorded at the Federal Aviation Administration Air Traffic Control Tower. Asheville Municipal Airport, Arden, July 19, 1967.
Transcript of Recorder Channel Number Two Approach Control/Flight Data Position. Asheville, NC, Department of Transportation, 1967.
Transcript of Ground Control Position. Charlotte, NC, Department of Transportation, 1967.
Transcription of Recorder Channel Number One Local/Ground Control Position Airport Traffic Control Tower. Asheville, NC, Department of Transportation, July 19, 1967.
"United Airlines, Inc., Boeing 727, N7030U, Salt Lake City, Utah, November 11, 1965." Aircraft Accident Report. Washington, D.C.: Civil Aeronautics Board, 1966.
"United Airlines, Inc., B-727, N 7036U, In Lake Michigan." Aircraft Accident Report. Washington, D.C.: National Transportation Safety Board, 1967.
"United Air Lines, Inc., DC-8, N 8013U, and Trans World Airlines, Inc., Constellation 1049A, N 6907C, Near Staten Island, New York, December 16, 1960." *Aircraft Accident Report.* Civil Aeronautics Board: Washington D.C., 1962.
"United States-Vietnam Relations: 1945–1967, Volume 1, The Air War in Vietnam." Office of the Secretary of Defense, Vietnam Task Force, 1971.
Van Sicklin, N.H. "Witness Statement of Mid-Air Collision," Hendersonville, July 21, 1967.
Walhout, Gerrit. *Human Factors Group Chairman's Factual Report.* National Transportation Safety Board, 1967.
Watts, R.H. "Accident Reports: Development and Use, Collection, Recording, Retrieval and Dissemination." Montreal, Session 7, Undated.
Weston, William. *Group Chairman's Factual Report.* National Transportation Safety Board, 1967.
Wilson, Mrs. Robert C. "Witness Statement of Mid-Air Collision." Hendersonville, July 20, 1967.
Woody, Richard. *Memo.* Meteorologist, WBAS, Asheville, NC, to Director, Weather Bureau, Silver Spring, Maryland, July 24, 1967.
Young, Warren R. "Turbulence-Hidden Giant in the Sky." *Life*, December 18, 1964.
"Youngsters Hunt Shelter." *Springfield Leader & Press*, July 20, 1967.

Index

Numbers in **_bold italics_** indicate pages with illustrations.

A & H Airport 21, 51, 152
Addison, David **_14_**–16, 24, 26, 27, 60–61, 74–75, 77–78, 88, 145, 154, 168, 174, 178, 182, 185; confusion of flight path 31–32, 48–49, 113, 177; flight training 12–13; garbled transmission 45–46, 146–147; partial exoneration 175–176, 183–184
Air Force Academy 93–94
Air Line Pilots Association (ALPA) 71–73, 112, 122–123, 132
Air Traffic Control Association (ATCA) 27, 72, 113, 117, 126
Air Transport Association (ATA) 72, 113, 122
Aircraft Owners & Pilots Association (AOPA) 24, 73, 78, 113, 116, 122, 132–133, 167
Airline Electronic Engineering Committee 111
airport congestion task force 27, 91–92, 99, 115–116, 118
airspace 16, 88, 98, 118, 125, 146, 148, 170, 176
All Nippon Airways 36
Allen, Bobbie 87, 100, 132, 139, 147, 157, 169, 171; House testimony 80, 82–86; _LIFE_ magazine article 42; _Washington Merry-Go-Round_ article 133–134
Allentown-Bethlehem-Easton Airport 100
American Airlines 35, 138, 149, 182
American Civil War 50, 102
American Revolution 50
Anderson, Jack 133–134, 159
Anderson, Robert 12–13, 45, 74, 77
Asheville, NC 12, 21, 24, 32, 41, 46, 50–51, 60, 71, 152, 167
Asheville airport (AVL) 7, 10, 13–17, 25, 31, 47–48, 51, 54, 60–62, 64–67, 74, 89, 98, 100, 105, 112, 124, 145–146, 153, 160, 168, 172–173, 175–177
Asheville Flying Service 64
Asheville Radio Beacon 26, 31–32, 45–46, 48, 60, 64, 145–147, 168, 176–177

ATC group 73, 113, 136, 139, 148, 160
Atlanta 7–8, 17–18, 25, 32, 141, 154
Atlanta Air Route Traffic Control Center 15–16, 23–24, 26, 31, 45–46, 61, 63–66, 145–147, 160, 176–177, 183
Automatic direction finder (ADF) 31, 48, 146, 153, 176, 184

Baltimore Airport 91
Barnes, J. J. 60–61, 65, 173
Beddingfield, Roy 55
Beechcraft (aircraft) 4, 28, 60, 82, 117
Bell, Bennett 113
Benham, Webster 17–18
Biltmore 50
Bishop, Ralph 62
Blossburg, PA 68, 82
Blue Ridge School for Boys 59
Blue Star Camp 52
Bodiford, Byron 112
Boeing Company 8, 33, 36–37, 40, 71, 73, 78, 83–84, 86, 113, 123, 132, 136–138, 157, 160–161, 168, 173–174
Boeing Factory School 11, 139
Boeing 707 9, 27
Boeing 727 7, 11, 23, 26, 27, 35, 37, 44, 46, 48–49, 56, 58–59, 63, 78, 84, 86, 91, 95–96, 110, 113–114, 132, 138–144, 149, 154, 167, 169–170, 173, 181, 184; special study of flight characteristics 34, 41, 47, 69, 80, 99–100, 175; Supersonic Transport (SST) 95–97; _see also Manhattan Pacemaker_; N68650
Boeing 737 8, 23
Boyd, Alan 99, 107, 117–118, 162–165, 170; Airport congestion (blue ribbon) task force 115–118
Braniff 38
British BAC 111, 4, 68
British Trident airliner 27
Broad River Radio Beacon 26, 31–32, 45–46, 48, 145–147, 153, 168, 176–177, 183
Brown, Evan 24, 45, 47, 60–67

209

210 Index

Bucks Elbow Mountain 22
budgets 102, 119, 125, 163–165; Defense 104; Department of Transportation 115; FAA 76, 93, 98–99, 105, 162; NTSB 27, 70
Buffalo, TX 38
Bureau of the Budget 27, 100, 101, 107
Burnett, Edwin 147–148
Burton, Clifford 27–28, 126

Camp Pinewood 52, *53, 55–57*, 58–60, 68
Cape Canaveral 120
Cartwright, H. M. 112
Cato, Earl 153, 175
Cessna 310 13, *14*, 15, 24, 45, 47–49, 59–61, 63–64, 67, 73–74, 78, 83–84, 136, 146–148, 150, 154, 160, 168, 170, 172, 174–177, 183; *see also* N3121S
Chaffee, Roger 120–121
Charlotte, NC 12–13, 15–16, 22, 74, 179
Charlottesville, VA 22
Cherokee Indians 50
Chicago, IL 91
Chidsey, James 17
cigarette 47, 49, 84, 113, 145, 148, 150–151, 168–169, 174–175, 180, 182, 184–185
Cincinnati, OH 21, 35, 37, 100, 149
Civil Aeronautics Board (CAB) 5, 22–23, 35–42, 47, 68–69, 77, 81–82, 88, 99–100, 110–111, 115–116, 139–141, 14–151, 156, 158, 167, 171, 174
claims 12, 71
Clark, Evert 163
Clausen, Clinton 70
cockpit vigilance 34, 149, 151, 174
Cockpit Voice Recorder (CVR) 3, 35, 84, 110–111, 113–114, 148, 150–151, 168, 175, 180
Collision avoidance system 90, 122–123, 148, 173
Colorado Springs, CO 93
Committee on Interstate and Foreign Commerce 78
Concorde 94
conflict of interest 70–72, 182, 184–185
Conrad, Thomas 7–8, 11, 25–26, 32–34, 46–49, 150
contributing factors 170
control tower 29, 45, 89, 124, 127, 153, 176; Asheville 7, 24–25, 32–33, 47–48, 51, 60–67, 74, 83, 145, 154, 173, 175, 184; Charlotte 13; shortage of 92–93, 98–99, 107, 117, 124, 126; Smith-Reynolds 10
Cox, Sandra 11
Crowe, Guy 44
Curtis, Bernie 147–148
Curtis, Kenneth 70
Curtiss Wright 95–96

Dacy, Ken 74, 78
Dampier, Ralph 112
Dando, Joe 12, 77–78

Daniel, James 113
Davis, Deborah 11
Davis, Tom 22–23, 71–72
Davis, Verne 53–54
DC-3 11, 21–22, 27
DC-4 65
DC-7 9
DC-8 4, 9, 27, 42, 88, 106, 119
DC-9 4, 28, 82, 179
Delta Airlines 68, 78, 82
Department of Transportation (DOT) 81, 99; budget 70; creation 69, 71, 115
departure procedures 47, 112, 136, 143, 152–153, 155, 167–168, 175
Dermid, Joel 54
Devine, Samuel 122–123, 128–130
Douglas Aircraft Company 40

Eastern Airlines flight 212, 179
Egypt Air 900 182
Electronics magazine 108, 127
Elkton, MD 42
Executive Order 109 88, 125–126

Fairchild-27 11
Fayetteville, NC 60
Federal Aviation Administration (FAA) 10, 27, 29–30, 35, 37, 39, 42, 44, 47–48, 64, 70, 72, 74–75, 77, 79, 81, 83, 87–91, 108, 111–113, 116–117, 132–133, 136–137, 142–143, 146–147, 152–154, 167, 170, 174–177, 179; budge 99–101, 105, 162–165; collision avoidance and 122; concerns about 727 36–37, 82, 149; concerns about Electra L-188 38; cutbacks by 43, 106–107; criticism of 109, 117, 124–131; placement of 69, 115; sonic booms and 95–96; SST and 93, 95, 97–98
Federal Bureau of Investigation (FBI) 74, 147–148
Federal Employees Veterans Association (FEVA) 125
Flight Data Recorder (FDR) 84, 110–111
Flight Recorder Group 73, 138–139, 159
Flight Safety Program 140–141
Florida Everglades 42
Folk, Benjamin 13–14
Forrestal 135
Fox, Leon 23
Friedel, William 91–92, 128
Friendship Airport 91
Fulbright, J. William 97

"G" forces 159
Gaither, Lee 112
General Aviation News 126
Gillilland, Whitney 68–69
Grand Canyon Crash 1956, 28, 88
Graves, Francis 112–114, 138, 145–148, 151, 178
Great Society 28, 71, 75, 162
Greenbrier Hotel 17–18

Greene, Joe 77
Greenville, SC 24
Griffin, Lacy 64–65
Grissom, Gus 120, 121
Grove Park Inn 167
Gulf of Tonkin 76

Haneda International Airport 36
Hartke, Vance 35–37, 40
Hawkins, Verl 15–16
Hendersonville, NC 50–52, 59–60, 62–64, 67, 72, 80–81, 83, *85*, 87–88, 90–91, 110, 134, 152, 171
Hogue, Prater 157, 160–161
Holiday Inn 52–*53*, 62–63, 66–67, 74
Howard Loadstar 63
Human Factors Group 72, 84–85

Instrument Flight Rules (IFR) 13–14, 32, 42, 46, 89, 127, 143, 153–154, 170, 174–175
Instrument Landing System (ILS) 15, 25, 31–32, 45, 48, 51, 107, 142, 145–146, 153, 176–177, 183
Interstate 26, 47–48, 51–*53*, *55–56*, 66
Investigator-in-Charge (IIC) 72, 84, 151, 156–158; *see also* Saunders, Thomas
Iran National Airlines 8, 136

Japan Air Lines 69
Johnson, Lyndon 19–20, 28, 35, 68, 70, 76, 99, *103*, 115, 117–118, 122, 162–164, 170; airport congestion and 116, 131; budget cuts 28, 75, 127; FAA and 27, 69, 96–97; hiring of additional air traffic controllers 165; McNaughton death statement 75; NTSB and 69, 71, 76, 81; Space race 118–119, 121; SST 93–96; Vietnam and 18–19, 29, 51, 75, 101–105, 165, 169
Johnson City, TN 21
Jones, Roys 24–25, 113

Kaufman, Bob 56–*57*, 58–59
Kennedy, John Fitzgerald 18–19, 93–95, 97, 106, 115, 118–119, 121, 125–126, 163, 170
Kennedy, Robert 97
Kennedy Airport 32
King, Martin Luther 170
Kuykendall, Dan 129–130

L-1049 9, 65; *see also* Super Constellation
Lake Central Airlines 68–69, 78, 82
Lake Michigan 35
Lake Pontchartrain 42
Lance, Beulah 18
Lanseair 12–13, 77, 172
Laurel, Oscar 69–70
LaViers, Harry 52–*53*, *56*
Lavrinc, George 21–22
liability 72- 78, 171–172, 181
Lockheed Electra L-1 88, 37–41

Love, William, family 18
Lyman, Stanley 77, 125–126, 131; call for McKee resignation 122, 125; House testimony 126–130
Lyons, Ken 77, 122, 125–126

MacArthur, Gen. Douglas 133
Malott, C. R. 113
Manhattan Pacemaker **9**, 48, 68, **85**, 145, 147; collision with vehicle 10, 41; leased by Piedmont 8, 23; problem with 41, 44; radar issues 42, 43, 124, 138; *see also* N68650
Marseilles, OH 68–69, 82
Martin 202 11
Martin 404 11, 23, 25–26, 32, 48, 51, 60–61, 145
McAdams, Francis 134, 176; cites Watkins actions 176–177, 183; confirmation hearing 69; leads public hearing 134, 167, 169; probable cause statement and 170–172
McKee, William "Bozo" 28–30, 35, 76, 88, 105, 124, 131, 164, 170–171; Airport congestion task force and 92, 99, 115–117; aviation budget 93, 101, 163–165; congressional testimony of 80–81, 87, 90–91, 98, 106–109, 162; cutbacks in FAA 27, 100, 127; Lyndon Johnson and 75–76, 79, 101; NAGE and 77, 117–118, 122, 125–126; SST and 96–98
McNamara, Robert 18–20, 75–76, 79, 102, 105, 170
McNames, L. W. 142–144
McNaughton, John T. 19, 76, 91, 102–105, 163; eulogized by Lyndon Johnson 75; memos and 103, 105; Pentagon Papers and 18–20, 104
McNaughton, Sarah 19, 105
McNaughton, Theodore 19, 105, 208
Merchant Marine 70
Meyer, Oscar 67
Meyer Flying Service 62
Miami International Airport 42
Miller, C. O. 171–172
Mitchell, Margaret 21
Mohawk Airlines 82
Monroney, Mike 95–96, 115
Mooney aircraft 60
Murphy, Charles 9, 116–117
Murray, William 56

N3121S 13, 45, 60–62, 64, 65; *see also* Cessna 310
N68650 8, 10, 41, 44, 123–124, 136, 138–139; *see also* Manhattan Pacemaker
Nashville, TN 60–61, 65
National Aeronautics and Space Administration (NASA) 40, 120, 174
National Association of Government Employees (NAGE) 29, 77, 79, 117–118, 122, 125–126, 131

212 Index

National Bureau of Standards 147–148
National League of Insured Savings and Loan Association 12
National Transportation Safety Board (NTSB) 41, 68, 70–71, 74–83, 88, 110, 115, 138–139, 144, 146, 165, 172, 176, 179, 180, 182; conflicts of interest 132–134, 158–161; final report 169–175, 177, 181; formation 69; investigative techniques 110–111, 113, 134–135, 147–148; party system 72, 84, 112, 120, 136, 156–158; petition of modification 182–185; public hearing 166–168
Nelmes, Edwin 113, 139, 143–144
New Orleans, LA 42, 68, 82
New York City Crash 1960, 88, 106, 119
New York Times 162, 164
Nicholson, Frank 113, 139–140, 142–143, 160–161
Nitze, Paul 19
Nixon, Richard 170, 178
Norfolk, VA 22
North American Aviation 95–96
Northwest Orient Airlines 38, 42

O'Connell, Joseph 69–70, 76–78, 80–82, 134, 172, 176
O'Hare airport 34–35
Office of Aviation Safety 82
O'Neill, Paul 40–41
Operations Group 73, 84, 113, 139, 144
Orrs Camp Road 59
Ottinger, Richard 106–109, 122, 131

Pacific Southwest Airlines 180
Pahl, John 39
Painter, Ronald **14**
Pan Am World Airways 42, 94
party system 72, 84, 112, 120, 136, 156–158
Patton, Ed 110–112, 159
Patton, Gen. George 133
Peace Corps 106
Pearson, Drew 78, 133
Pekin, IL 18
Pentagon Papers 18–20, 104
Pickle, J. J. 99–100
Piedmont Airlines 7–8, 12, 23, 25, 44, 52, 72–74, 78, 81–84, 86, 110–113, 134, 140, 144, 157–160, 168, 170, 172, 174; fleet 9, 23, 137; formation of 11, 21, 51; management and 41, 43–44, 71, 114, 141, 183–184; safety and 41, 43–44, 123–124, 137–140, 142, 149–150, 169
Port, Clifford 62, 66
Powers, Kitty 57–58
Pratt & Whitney 8, 78, 95–96
probable cause statement 71, 78, 81, 83, 167, 170–172, 176–177, 179, 181
Project Apollo 120–121
Project Gemini 120
Project Mercury 120

Proxmire, William 97
public hearing 37, 68, 81, 82, 109, 134, 139, 146, 148, 151, 166–167, 169

Quesada, Elwood "Pete" 39–41

racial unrest 51–52, 164
radar 10, 15–16, 63, 70, 82, 85, 89–90, 92–93, 97–101, 106–107, 120, 124, 128–129; AVL and 26, 45, 51, 89, 105, 155, 176, 183; N68650 and 42, 44, 86, 112, 124, 136–139, 173; weather and 23, 43, 91, 123, 125
RAND Study 158
Rapid Air 12
Rawson, John 112, 136–137
Rayburn House Office Building 80
Reagan, John 111, 124, 138, 159–161
Reed, John 69–70, 74–75, 77, 83, 134
Republican 70, 115
Reynolds, Ralph 12–13, 77
Roanoke Airport 22, 24, 33, 45, 153–154
Roberts, Harold 64–66, 74, 78, 83
Rockford, IL 89
Rogers, Paul 92–93, 97–99, 106, 116–117
Rooney, Fred 100–101, 105
Roosevelt, Franklin 77, 133
Ruby, Charles 122–123

Safety Recommendation A-74-85 179
St. Louis, MO 27
Salt Lake City, UT 35
Saunders, H. K. "Zeke" 21, 71; appointment of Piedmont party representatives 73, 158–159, 184; conflict of interest 133–134, 159–161; deposition regarding duties 72; meets with Piedmont party representatives 159–160
Saunders, Tom 72, 74, 83, 112, 132, 137, 149; appoints parties to investigation 84, 157, 159–160; conflict of interest 133–134, 160–161, 182–183; Investigator in Charge (IIC) 72, 151, 156, 177; 727 special study and 100, 113–114, 135, 160, 169
Schulte, Ray 7, 8, 10, 12, 24–26, 32–34, 46–48, 89, 153–155, 159, 168; cigarette and 49, 150–151, 169, 182, 184; 727 training and 133, 141–144, 174
Schultze, Charles 116, 162–165
Shriver, Sargent 106
Shuler, Martin 18
Smith Reynolds Airport 10, 41, 44
Snell, Paul 25–26, 32–33, 46–48, 51, 60–61, 63–64, 66, 145, 154, 168
sonic boom tests 94–96
Soviet Union 19, 118–119
Space race 118–120, 127
Springfield, MO 12–13, 74
Staggers, Harley 78–81, 83, 87, 90–91, 109
Sterile Cockpit 179–180
Stokely 17

Structures Group 72, 84, 86
Subcommittee on Transportation and Aeronautics 78, 122
Supersonic Transport (SST) 93–98

Tadlock, W. O. 149–150, 160, 168
Tell City, IN 38
terrain avoidance 153, 175
Thayer, Rear Admiral Louis 69–70, 82
Thomas, David 80, 87–90, 92–93, 98, 126
Timmerman, Craig 113
Tipton, Stuart 124–125
Tokyo Bay 34, 36
Trail of Tears 50
Trans World Airways (TWA) 28, 82, 106, 117, 119, 182
Trippe, Juan 94
Truman, Harry 19, 69, 77
turbulence 40, 42, 82–83, 132, 159; *see also* weather

United Air Lines 28, 35–36, 106
United States Coast Guard 70
United States Congress 6, 28, 76–77, 80, 87–88, 99, 107, 115–117, 126, 130–131; aviation funds and 92, 96, 98, 100, 156, 162–163; aviation safety hearing 6, 28, 76–77, 80, 87–88, 99, 107, 115–117, 126, 130, 131; McKee and 27; NTSB and 69, 72, 156, 171; 727 safety issues and 10, 34–37, 86; SST and 97
Urbana, OH 28–30, 68, 82, 87–88, 91, 122, 126

Valdese Intersection 24–25, 49, 154
Van Sicklin, N. H. 62–63, 66–67
Very High Frequency Omni-directional Radio range (VOR) 25; Asheville (Sugarloaf Mountain) 14–15, 26, 31–33, 45–48, 60, 145–147, 154–155, 176–178, 183–185; Spartanburg 24

Vietnam 18, 76, 101, 165, 169; body count 29; bombing of North 19, 76, 104; funding of 28, 71, 75, 79, 104; Lyndon B. Johnson and 79, 102, 102, 104; Pentagon papers 19; unpopularity of 19, 51, 105, 118, 135, 164, 178; *see also* McNaughton, John T.
Visual Flight Rules (VFR) 13, *14*, 42, 46

Walker, Clifford 36
Washington, D.C. 7–8, 10, 18, 21–22, 29, 36, 68, 91, 110, 134, 159, 170
"Washington Merry-Go-Round" 78, 133; *see also* Anderson, Jack; Allen, Bobbie
Washington National Airport 83, 91
Watkins, James 23, 27, 31, 45, 47–48, 63, 65, 178, 182–184; attempting to locate N3121S 60–62, 64, 66; cited by McAdams 177; mistakes by 146, 168, 175–176; receives "garbled" transmission 45–46, 136, 145, 147; sends wrong instructions to N3121S 26, 31, 46, 48, 74, 113
Watson, Albert 99–100
Watson, L. A. 124
weather 13, 36, 42–43, 46, 51, 54–55, 72, 88–89, 98, 118, 163, 173, 175; *see also* turbulence
Weather Bureau 43, 76, 172
Weather Group 73, 84–**85**
Weather Radar 13, 36, 42–43, 46, 51, 54–55, 72, 88–89, 98, 118, 163, 173, 175
Welch, James 24–26, 32–33, 60, 63, 154
White, Ed 120–121
Wilmington, NC 21
Wilson, Lawrence 11, 25–26, 34, 49, 150–151
Wilson, Mrs. Robert 54–55
Winston-Salem, NC 8, 10, 41, 44, 71, 112, 141, 159, 174

Yarmouth Castle 70

www.ingramcontent.com/pod-product-compliance
Ingram Content Group UK Ltd.
Pitfield, Milton Keynes, MK11 3LW, UK
UKHW041957140426
5217IPUK00015B/848